# 99 Ways to Improve Your CB Radio

by LEN BUCKWALTER

HOWARD W. SAMS & CO., INC.
THE BOBBS-MERRILL CO., INC.
INDIANAPOLIS · KANSAS CITY · NEW YORK

# Preface

Whether a CB rig is rolling over the road or setting at home at a base station, time imposes its toll on the like-new condition. Components age and change value, dust accumulates, and corrosion often sets in. Curing these problems is generally easy, and many troubles can be prevented from occurring.

The tips and procedures in this book have been tested and proven on Citizens band equipment. They can be used to keep your CB rig in top operating condition. Little technical skill is needed to perform the work, and all of the adjustments described here remain within the limits set by the Federal Communications Commission.

You will also find many hints on operating techniques, installation techniques, and treatment of noise interference. Many practical tips on station setup can improve the appearance as well as the operating efficiency of your base station.

All of these various tips, techniques, and procedures help support the original concept behind CB radio—that it can be a low-cost, two-way service within the reach of anyone.

LEN BUCKWALTER

# Contents

## SECTION 1.  Antennas

## SECTION 2.  Interference Suppression

## SECTION 3.   Equipment Maintenance

## SECTION 4.   Accessories

## SECTION 5.  Operating Techniques and Aids

## SECTION 6.  Additional Information

*SECTION 1*

# Antennas

# 1

## Antenna Height

Since the height of a CB antenna is closely regulated and highly visible, let's consider key requirements which affect installation. There are several considerations.

Most important is the rule which states that an antenna and its supporting structure must not rise higher than 20 feet above the mounting point. The mounting point is considered to be a natural formation, tree, or man-made structure. In an earlier version of CB regulations there was confusion about what constituted a man-made structure. Some operators believed that a tower, mast, or pole are man-made structures and could be used to support a CB antenna. Later rules, however, clarify this point and eliminate these mounting possibilities.

An exception occurs in the case of a tower being used for the antenna of a transmitter already licensed in another service (broadcast, etc.). Here the CB antenna may be mounted on the tower, but may not exceed the tower height or 60 ft. above ground level.

There is also a regulation which applies when the CB antenna is mounted on an antenna structure which is being used only for *receiving*, a television antenna for example. This is permissible, but only if the receiving antenna structure is not more than 20 feet higher than a natural or man-made structure on which it is mounted. If the antenna is omnidirectional, the highest point of the antenna and its supporting structure may not exceed 60 ft. above ground level.

# 2

## Antenna Guy Wires

For supporting a heavy antenna, or when using a mast much over five feet in length, guy wires are frequently required for additional support. Consider this hint for proper guy-wire performance.

Guy wires should not be pulled absolutely tight during installation. A slight amount of slack is advisable to prevent breakage due to antenna movement in the wind or contraction of the wire during cold weather. Use weather-resistant wire of the stranded, galvanized type. Carefully avoid any kinking of the wire during installation.

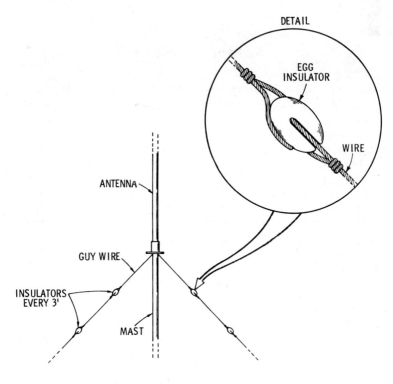

Fig. 2-1. Method of installing guy wires.

Since guy wires are also conductors of electricity, it is extremely important to avoid any lengths which could possibly pick up or tune the antenna signal. The result is a loss of power and distortion of the antenna's radiation pattern. The method for preventing this is shown in Fig. 2-1. Each guy wire is broken up by placing an insulator every three feet along its length. This prevents any accidental antenna effect since three feet of wire will not respond to the 27-MHz signal.

For reasons of safety, it is important to use only egg-type strain insulators for breaking up guy wires. As shown in the detail, a broken insulator will not cause separation of the guy wire; there is still support. A broken insulator, of course, should be replaced.

# 3

## Reconditioning Antenna Elements

The ravages of weather, corrosive fumes, and age may reduce the efficiency of CB antenna elements. A once-a-year treatment can keep antennas operating effectively and help stave off problems before they occur.

Soot and other deposits can build up on plastic insulators and create a short-circuit path for energy between, say, a vertical element and radial rods below it. A vigorous rubbing with cloth is usually enough to restore insulators to a like-new condition.

In telescoping elements of an antenna, where elements have been slid together, there is a potential loss of efficiency due to oxidation between the rods. Disconnect the elements and restore their mating surfaces by cleaning them with steel wool. Best results occur if the ends of the elements are then coated with conductive grease (available at radio supply stores)

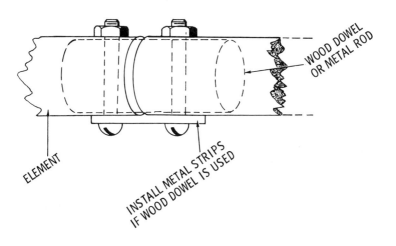

**Fig. 3-1. Repairing broken antenna elements.**

before sliding them together. This ensures an excellent electrical path for long periods of time. After elements are reassembled, it's a good idea to cover the joint with an acrylic spray to prevent any entry of moisture.

Elements which have become damaged, either bent or broken off, can be repaired by the technique shown in Fig. 3-1. It requires the insertion of a wood dowel or aluminum tubing (or rod) between the broken elements. If metal tubing is used, the required electrical connection occurs when the assembly is bolted together. When you are using wood dowel, however, the electrical connection will have to be made by running some metal strips to provide good continuity between elements.

# Preferred Antenna Position

The choice of an antenna location on a car is influenced by such factors as ease of installation, cutting holes in the car body, and overhead obstructions presented by a garage. But in terms of electrical performance, there are five basic locations, as illustrated in Fig. 4-1. Consider each of these positions in order of preference:

1. Roof-top. Placed squarely in the center of the car roof, an antenna in this position provides the most signal strength in each direction. Since the roof metal is most equally distributed around the base of the antenna, the pattern is circular or nondirectional. There will be little or no "beam" effect. Even if the car is driven in a circle, a distant receiving station hears no difference in the signal level. Another factor in favor of the roof-top mount is that it permits the signal to clear nearby obstructions. Also, there is no blocking of the signal by any other metal surface of the car.

2. Trunk Mounting. Somewhat less antenna efficiency occurs when the whip is positioned in the center of the trunk lid; for this reason, it is the second-best location. The result of having more metal in front of the antenna produces a small increase in power toward the front of the car (in the direction of travel). This increase is at the expense of power transmitted toward the rear of the car.

3. Fender Mounting. This position is on the rear fender, right or left side, next to the rear window. If the right fender is used, the strongest signal is transmitted in the direction of the *left front* fender. For rear left mounting, the signal is strongest toward the right front fender. The reason for these differences is that the signal tends to be drawn over the longest metal path

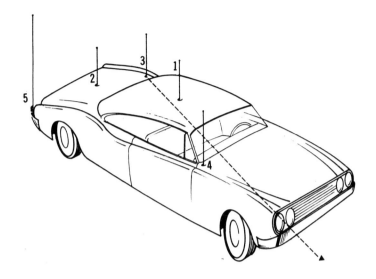

Fig. 4-1. Preferred antenna positions.

presented by the car. This is pointed out for the antenna marked 3 in Fig. 4-1. Note how the signal runs diagonally across the car roof in its favored direction. There is a corresponding drop in strength in the opposite direction. These effects are most noticeable at longer ranges. This should be taken into account during an emergency call. By turning the car and taking advantage of these directional effects, it might be possible to get the message through under poor-signal conditions.

4. This mounting point, on the front cowl (either side), produces the same general effects as already described for the rear-fender mount. Now the favored direction is toward the rear of the car, since the largest mass of metal lies in that direction. A similar diagonal effect also occurs; a left front mounting produces strongest signal toward the right rear of the car.

5. This last case is the bumper mount at the rear. Although this can be the easiest point for mounting the whip, it has certain disadvantages. For one, it places the base of the antenna, where most of the signal exists, extremely close to the trunk. This causes some obstruction in the forward direction. But since the forward direction also favors the signal (most metal lies in this direction) there is also some strengthening in that direction. The net result is that a rear left bumper antenna produces the best signal toward the right front fender; the right bumper mount favors the left front. Another factor which works against the bumper mount is that maximum signal power is placed quite low on the car where it may not successfully clear obstructions close to the car.

15

# 5

## Mobile Roof Mounting

Of the possible mounting points for a car antenna, it is generally agreed that the center of the roof out-performs all other locations. Not only does it raise the antenna to the high point of the car, but it also improves the needed ground-plane action of the metal roof. It is somewhat more difficult to install an antenna at this location, but here are tips which could simplify the job.

The usual run for the coaxial cable is under the metal roof and above the headliner (over your head as you sit in the car). The cable starts at the antenna mounting point, then runs to one of the front posts that separates the windshield from a side window. Several of these details are illustrated in Fig 5-1.

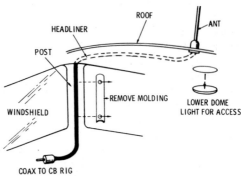

Fig. 5-1. Mobile roof mounting.

To determine whether the job is practical, unscrew the molding from the window post. (Either side will do.) Then run a heavy piece of wire from a post to center of the car roof. If you can snake through to this point, the job of routing the coaxial cable should be relatively simple. Next, check on how the hole in the car roof will be made. If the car has a dome light near the center of the car roof, the fixture can be dropped for access to the underside of the roof. Otherwise, you might have to lower part of the headliner. Since this is not easy to do in some cars, try to avoid this by using an antenna that mounts from the top and needs no nut-tightening at the underside of the roof.

Drilling the roof hole should be done with extreme care. Distribute your weight evenly to avoid denting the roof with your shoes. Hammer a small dimple in the roof with a nail or center punch, so that the drill will not slip. Use very gentle pressure on the drill or otherwise the spinning bit will cut through the thin sheet metal and cut into the headliner. Placing a wood block next to the bit can prevent this. Once the drill hole is made, it is widened to the required diameter by a hole saw or a tube-socket punch. You could file it to size, but this takes much longer.

By taping one end of the coax cable to a stiff wire (or a coat hanger) it should be possible to snake the cable between the headliner and roof to the window post and down to the CB transceiver.

Another way to win the benefits of a mobile roof-top antenna, but without cutting a hole in the car, is with a magnetic mount (Fig. 5-2). The model shown has a powerful permanent magnet in its base for an instant grip on the roof. It can withstand high wind forces while driving. An added feature in some models is an optional adhesive to form a near-permanent bond with the surface. With a magnetic mount you not only install an antenna without holes or routing a cable through the headliner, but can change positions on the car or remove the antenna quickly. One disadvantage to consider: an exposed cable will run from the antenna across the car roof, and it must enter the passenger compartment through a slightly open window.

**Fig. 5-2. Magnetic mount antenna.**

# Whip Adjustment

There is some compromise built into some mobile whip antennas. This is due to the fact that the manufacturer cannot know in advance exactly where on the car the whip will be mounted. Different locations yield different results. Placing an antenna on a bumper might position it inches from the trunk. A whip atop the roof "sees" a different pattern of metal. Since the metal area serves as an electrical ground, its effect will vary with antenna location and the shape of the car.

One result can be an increase in the antenna's swr (standing-wave ratio), which steals power by reflecting it back to the transmitter. This can be checked by inserting an swr meter in the coax line leading to the whip. In some mobile antennas there is provision for adjusting the swr to the desired low value, but in others, some "pruning" is needed to reduce the swr reading.

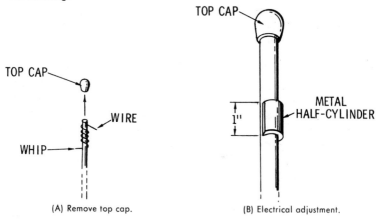

(A) Remove top cap.  (B) Electrical adjustment.

**Fig. 6-1. Whip adjustment.**

In most cases, swr is higher than normal when the whip is near metal surfaces which reduce the whip's resonant frequency below 27 MHz. One method for determining this is to move your hand toward the antenna while the swr meter is operating. If this *increases* the swr reading, chances are that the antenna requires shortening (to raise its resonant frequency). This can be done to spiral-wound whips by first removing the top plastic cap. (See Fig. 6-14A.) Grasp the wire and unwind one turn only, then cut it off, and check for a lowering of swr. This process, cutting off one turn at a time, may be continued until the swr reading is below about 2 to 1. Be careful not to go too far since it may not be practical to restore missing turns.

If the whip is indicating high swr due to insufficient length, this is usually indicated by a *lowering* of the swr reading as you bring your hand near the operating antenna. When there is no practical means for physically lengthening the whip, as in the case of a spiral-wound type, it is possible to do it electrically (Fig. 6-1B). This is done with a 1-inch piece of metal tubing that has the approximate diameter of the whip. The tube is sawed in half lengthwise, so you end up with two half-cylinders. Using just one of these 1-inch half-cylinders, slide it up and down the top section of the antenna while watching the swr meter for the lowest reading. When it is found, fasten the tube permanently in place with tape or some other suitable adhesive.

# 7

## A Safe Antenna-Grounding System

Since lightning tends to strike the highest point in a given area, a CB antenna could be a likely target. The electrical path can occur from the antenna elements, through a coil at the antenna base, down the shield of the coaxial cable, then to the transceiver chassis. Unless the electrical path is continued to a ground other than the transceiver, there is a possibility of damage to components within the set. There are several ways to provide a ground for protection from lightning.

In the simplest grounding arrangement, a heavy wire is run from the transceiver case to a cold-water pipe or to the screw which holds down a cover plate of the a-c wall outlet. A more elaborate grounding system, however, greatly increases the safety factor. It also overcomes a possible difficulty in using a cold-water pipe. Plumbers frequently apply "dope" to pipe joints which can insulate the pipe from the ground. The a-c cover plate, too, may not be directly at electrical ground in some homes.

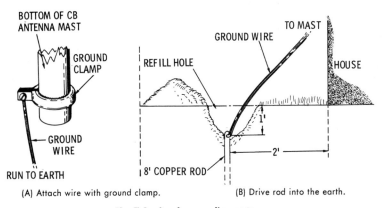

(A) Attach wire with ground clamp.    (B) Drive rod into the earth.

Fig. 7-1. A safe grounding system.

A really effective ground is neither expensive nor difficult to install in the instance of a private home. (The items are available where electrical supplies are sold.) First is a length of bare wire—No. 8 aluminum or No. 10 copper. One end of the wire is attached to the lower end of the antenna mast with a grounding clamp (Fig. 7-1A). The wire is then led down, as straight as possible, toward the ground below. Avoid unnecessary turns or loops.

The wire will be fastened to a ground rod after some preparation. The rod is a standard 8-foot copper rod driven directly into the earth (Fig. 7-1B). The rod must be located at least two feet away from the house. Furthermore, the top of the rod must be submerged *below* ground level by at least 1 foot. For this reason it is convenient to locate the point where the rod will be driven into the earth and first dig a hole slightly over a foot deep. Then, hammer the rod down until its top end is just visible in the hole. This permits the ground wire to be attached. (If the rod has no provision for fastening the wire, a grounding clamp should be used.) The last step is filling-in the hole and tamping down the earth until it is firm.

# Avoid Antenna Interaction

To provide for grounding when a base-station antenna is mounted at roof height, several radial rods usually jut outward from the lower elements. They provide an electrical ground that would otherwise be established by the earth. To function properly, these radials, however, must not interact with actual earth. To operate with efficiency, the radials should be no less than one-half wavelength above the earth. This is a minimum of 18 feet. At lower heights the radials tend to warp the antenna's full-circle pattern.

A beam antenna has no radial elements, but it is susceptible to interaction with nearby objects made of metal. To prevent losses, beam elements should lie at least 12 feet, more if possible, from such objects as metal rain-gutters, copper or aluminum roof flashing, and electrical cables used for power or telephone. Ample spacing not only reduces signal loss but also preserves the sharp, directional characteristics of the beam.

# 9

## Calculating Antenna Gain

Here is a simple chart designed to aid in the selection of an antenna. It converts an antenna's advertised dB gain into the increase of effective radiated power, in watts, that may be expected. The chart is based on a typical transmitter of 5-watt input and as an output power of 3.5 watts.

To use the chart, find the number of advertised dB in the left column. Then read the resulting power increase in the right column. For example, an antenna offering a 6-dB gain will increase its effective radiated power output from 3.5 watts to nearly 14 watts. (Note that for each 3 dB of gain, there is an approximate doubling in effective radiated power.)

| Antenna Gain | Power Output |
|---|---|
| 3 dB | 6.97 watts |
| 4 dB | 8.79 watts |
| 5 db | 11.06 watts |
| 6 dB | 13.93 watts |
| 7 dB | 19.67 watts |
| 8 dB | 22.09 watts |
| 9 dB | 26.79 watts |
| 10 dB | 35.00 watts |

## Concealed Mounting Holes

When it is time to trade in an automobile, some operators become concerned over the hole left by the CB antenna mounting. And drilling a hole in a new car may not be a good prospect either. There are several antenna-mounting possibilities which help reduce the loss in the car's resale value.

The use of a CB/a-m *splitter* permits the CB antenna to be mounted in the hole ordinarily used for the car's standard radio. This accessory permits the CB antenna to operate for both CB and the regular radio. Follow the splitter manufacturer's recommended installation carefully. To prevent any deterioration in the regular radio's performance, be sure to check the adjustment of the antenna trimmer. This adjustment is found on auto radios near where the antenna is plugged in or behind the tuning knob. A label at this point usually gives the procedure: to tune the radio to 1400 kHz and adjust the trimmer (with a small screwdriver) for maximum audio.

Another approach which prevents marring the car may be used when the auto has no back-up lights. The antenna base is mounted in a hole drilled at the rear of the car. When the car is ready to be sold, the antenna is removed and a back-up light installed in the same hole. This technique might also be used in a car already equipped with back-up lights. One of the lights is removed and its hole used for the antenna mount.

The popular roof-top mounting generally requires a very small hole which can be repaired at a future time. Before drilling the roof hole, you may wish to check with a local body-and-fender shop. Some estimates have shown that the job of repairing the hole is a minor one and can be done at little expense.

Another approach is the use of an antenna which functions for both CB and standard broadcast. This unit operates much like the splitter system mentioned earlier. Included is a special phasing harness which properly processes a-m and CB signals and channels them through suitable cables and plugs.

## Mobile Range Boost

Since many CB mobile antennas are not mounted on the center of the car roof, they are not completely nondirectional. That is, they tend to transmit signals more strongly in certain directions than in others. This may even occur in roof-top mounting due to unequal distribution of car metal around the base of the antenna. The general effect is to strengthen the signal which travels over the longest dimension of the car. An antenna which is positioned on the front left of the car, for example, tends to radiate best toward the right rear.

This inequality can be used to advantage in situations where it is important to obtain communications under poor conditions. Once the directivity of the antenna is known, it is possible to position the car accordingly. While this may not be practical on a routine basis, it could prove helpful during a roadside or other emergency.

To discover the best direction of the antenna, you can follow a simple procedure. It requires the assistance of a base station. Drive the car several miles from the base station until your S-reading, read at the base station drops below S-5. (At these levels, differences in signal are more readily seen on the meter.) The car should be in the clear, away from overhead power lines and obstacles. There should be sufficient open area to permit you to drive a complete circle, making a tight turn.

Begin by noting the starting point and call the base station and have it take an S-reading. Make the transmission about a half-minute long, to rule out the effects of signal flutter which may occur due to passing cars. Ask the base operator to give you an average S-reading for the half-minute period.

Next, drive the car slowly in a complete circle. Imagine the circle as divided in about eight points. Obtain readings at each point with half-minute transmissions. The complete turn will indicate the point, or points, of best signal radiation. For increased accuracy, the car should be driven to another location and the test repeated. This rules out reflections from hills and other obstructions which may give false readings. Once the result is known, you'll know how to point the car for optimum communications.

# 12

## Coaxial Cable Length

The coaxial cable running between transmitter and antenna is normally not considered a "tuned" element. It may be any length and still preserve its rating. In the case of RG/58U cable, commonly used for CB work, the impedance rating is 50 ohms (approximate). Whether the cable is cut to two feet or two hundred feet, it should retain this characteristic impedance.

In some instances, however, the cable will respond unfavorably in a particular antenna installation. This reduces the transmitter's ability to load power into the cable and ultimately to the antenna. One technique for improving the situation is to use a length of coaxial cable which is an electrical half-wavelength, or a multiple of this dimension. With RG/58U, an electrical half-wavelength occurs when the cable is cut to 11 feet 10 inches. (Note that this is shorter than a half-wave *antenna*, which is approximately 18 feet. The reason is that radio-frequency energy travels more slowly through cable than through air, and the cable must be correspondingly shorter to achieve the same effect.)

Thus in choosing a cable length for a particular antenna installation, mobile or base, cut it to 11 feet 10 inches or any multiple of this length, such as 23 feet 8 inches, 35 feet 6 inches, etc.

# 13

## Emergency Antenna

If, for some reason, you must put a signal on the air quickly and have no conventional antenna, it's possible to rig a temporary length of wire that will radiate a reasonable amount of power. It's done by taking a length of ordinary hookup or other wire cut to 108 inches. Form a tiny loop at one end and insert it into the transceiver's antenna jack. The other end is run straight up, as vertically as possible, and taped to a high molding or ceiling. Performance of the temporary antenna is improved if the transceiver has a good electrical ground; a wire from the chassis to a cold-water pipe, for example.

# 14

## Temporary Antenna

An easily constructed half-wave antenna might come in handy for temporary installation. The one described here takes up little space and may be rolled into a coil when not in use. Although the design is not as efficient as a regular base-station antenna, this may be outweighed by simplicity.

Construction details are given in Fig. 14-1. One end of a length of 72-

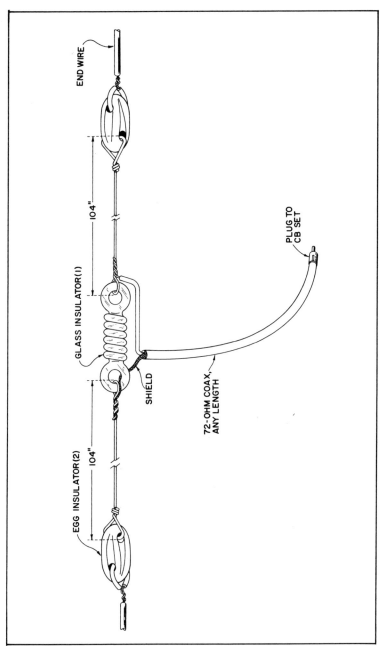

Fig. 14-1. Temporary antenna.

25

ohm coaxial cable is prepared so the shield and center wire may be connected to each wire arm of the dipole. Egg-type strain insulators at the outer ends permit rope or wire to fasten the antenna in place without short-circuiting the wire elements. Those wire elements, each 104 inches long, may be regular stranded antenna wire or No. 18 solid copper wire.

Since virtually all CB antennas are vertically polarized, it is advisable to mount this antenna vertically for best effect. This is done by tying one end of the antenna, using either end wire or rope, to a tree limb or other high support. Another improvement occurs if the coaxial cable is brought away from the middle of the antenna at right angles for a distance of at least eight feet. Bringing the cable down alongside the lower antenna wire is apt to introduce some energy losses.

# 15

## Telescoping Elements

The construction of most base-station antennas utilizes telescoping aluminum tubes. The tubes are generally fastened at each joint by a sheet-metal screw. But if the antenna is dismantled, you may discover that element sections have become corroded and are difficult to slide apart. This problem is eliminated by coating the mating surfaces with either conductive grease or silicon-type lubricant during installation of the antenna. Not only does this eliminate locked joints, but it can also reduce noise problems which sometimes occur when corrosion develops where the tube sections join.

*SECTION 2*

# Interference Suppression

# 16

## Spark-Plug Noise Suppression

This is the most common type of interference and frequently the strongest. It is caused by high-voltage sparks jumping across the gap of spark plugs and inside the distributor. The car should be equipped with at least one of the various techniques available for suppression.

Identifying spark interference is done by turning off the engine, then turning the ignition key to the accessory position, and listening to the CB transceiver. The ignition noise should disappear. When the engine is running, the speed of the popping noise keeps in step with engine rpm.

Before taking corrective measures, see if the car is already equipped with spark-noise suppression. This is done by examining the cables which run from each spark plug to the distributor. If they are marked with words like "radio resistance," this indicates that suppression measures are present (true for many late-model autos). Older cars can often be equipped with this special high-voltage resistance cable. Be sure to obtain a cable set intended for your car.

Another approach is the use of special resistor-type spark plugs. These units have built-in suppressor resistors. These are also available as separate units which plug into the top of each spark plug and into the center hole of the distributor cap.

Be sure that all high-voltage cabling is in good condition. Small cracks in the insulation are apt to cause electrical leakage and noise.

The most elaborate form of plug and distributor suppression is complete shielding. Although this is intended primarily for marine installations (since a boat's wood hull provides no shielding effect), there are kits available for automobile ignition systems. In most instances, however, ample noise reduction is possible with the simpler techniques mentioned earlier.

# 17

## Alternator Filter

In the mid-1960's there was a gradual changeover from the generator to the alternator in auto ignition systems. Not only does an alternator charge the battery at engine idling speeds, but it reduces the amount of noise picked up by the CB transceiver. Yet there are occasional cases where filtering of the alternator can produce quieter reception.

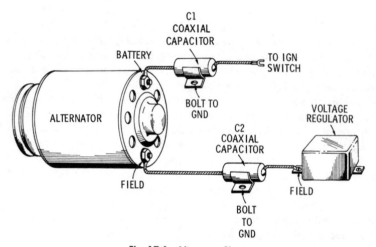

Fig. 17-1. Alternator filter.

Shown in Fig. 17-1 are two points in the system which may be bypassed with high-quality coaxial capacitors. These capacitors (such as Sprague 48P18) are rated at a capacity of 0.5 $\mu$F and current handling of 40 amperes. They operate by short-circuiting the offending noise to ground, while leaving dc power from the alternator unaffected. The mounting tab on capacitor C1 is bolted to the frame of the alternator. For a good electrical ground, the tab must contact a clean, oil-free surface.

Depending on the particular car, one or both capacitors are required. This can be checked by installing each capacitor and noting its effect on noise pickup in the receiver. Capacitor C1 can be installed at the alternator's output lead, which may be marked "Batt" (for battery). The lead is cut and connected to the end screws of the capacitor. The other capacitor, C2, may be connected at the voltage regulator's F, or field, terminal.

# 18

## Alternator Rings and Brushes

To cure noise interference from an automobile alternator, filtering is usually used. But, an additional step can help to further reduce alternator noise at the source. Cases of noise have been traced to dirty slip rings and worn brushes. Although these components produce less sparking than the brush-and-armature arrangement in the conventional car generator, they can still produce enough interference to reduce receiving range.

Whenever an alternator-equipped car is given an ignition tune-up—or at approximately 10,000-mile intervals—the serviceman should remove the alternator and open its housing. This permits cleaning of the two slip-ring surfaces. At the same time, brush length and spring tension should be checked. Some alternator manufacturers recommend that when brushes are worn down to one-half of their original length, replacement is advisable. Each of the items can produce sparking and the radiation of noise.

Another factor is an out-of-round condition of the slip rings in older alternators. Even a small error, as little as 0.0005 inch, may upset normal alternator operation, and possibly increase noise level. This, too, can be determined when the alternator is removed for inspection.

# 19

## Generator Whine

Sparking at the carbon brushes of a car generator is a frequent cause of CB interference. The symptom is a rough buzz or whining sound that appears to rise in pitch as engine speed is increased. It is independent of the car's speed over the road. In bad cases of generator noise it is advisable to have a new set of brushes installed, a remedy that may help the operation of the generator as well.

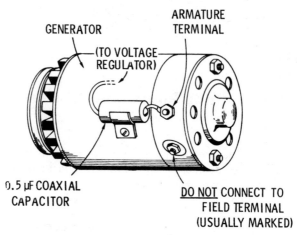

GENERATOR

ARMATURE TERMINAL

(TO VOLTAGE REGULATOR)

0.5 µF COAXIAL CAPACITOR

DO NOT CONNECT TO FIELD TERMINAL (USUALLY MARKED)

Fig. 19-1. Eliminating generator whine.

Suppressing interference is done with a coaxial capacitor of 0.5 µF (Fig. 19-1). The capacitor is installed at the generator's armature terminal, sometimes marked A. At this point there may be found a factory-installed capacitor. It is designed for reducing noise in the standard a-m broadcast band and has insufficient effect at 27 MHz. This capacitor is removed completely and the coaxial type is substituted.

The metal mounting lip of the coaxial capacitor must make good contact with ground, or the frame of the generator. Clean away grease or corrosion at this point. The generator armature terminal is then connected to one end of the coaxial capacitor, and the wire which is removed from the terminal is screwed to the other end of the capacitor.

# 20

## Regulator Noise

An irregular rasping noise that comes and goes can often be traced to the car's voltage regulator. Noise is developed by sparks created at the regulator's vibrating contacts. Reducing this interference can be done by installing a 0.25-$\mu$F coaxial capacitor at the regulator, as in Fig. 20-1.

For best effect, keep the capacitor as close to the regulator as possible. The mounting lip is bolted to the nearest metal ground. It is seen that the lead originally connected to the regulator battery terminal is removed and reconnected to one end of the capacitor. A short, heavy wire is then run from the other capacitor end to the regulator battery terminal.

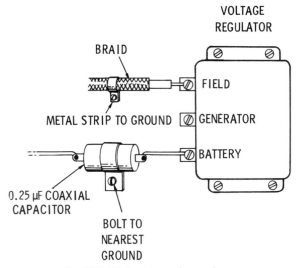

Fig. 20-1. Eliminating regulator noise.

In some cases, it may be necessary to also install an identical coaxial capacitor at the regulator generator terminal. Do not connect any capacitor to the field terminal. If the regulator still seems to be producing noise, try shielding the field lead, which runs to the generator. Hollow copper braid, available for the purpose, is pulled over the full length of the field lead. The braid is then connected to metal ground at both ends by metal clips. Be careful not to short-circuit the ends of the field lead.

## Ignition-Coil Noise

There is a possibility that noise energy at the car's ignition coil may find its way into the lead which brings battery power to the coil. By confining the noise to the coil, there is less chance of interference signals spreading over the entire low-voltage wiring system. Interference of this type is characterized by a popping noise heard in the CB set while the engine is operating. It varies with engine speed.

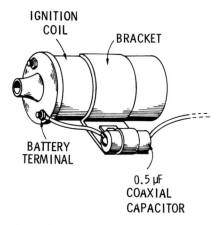

Fig. 21-1. Eliminating ignition-coil noise.

The method for reducing coil noise is by bypassing it with a coaxial capacitor. Use a 0.5-$\mu$F unit and install as shown in Fig. 21-1. The capacitor must mount as close as possible to the coil terminals. A bolt on the coil-mounting bracket may prove practical. Wiring the capacitor is done by removing the lead going to the coil's battery terminal, and connecting it to one end of the capacitor. A short, heavy wire is then installed between the other end of the capacitor and coil battery terminal.

# 22

## Gauge Interference

After the noise from the car's ignition system is suppressed, there may be noticed electrical interference from other sources. A typical source is the various gauges which create sparking at their operating elements and consequent radio noise. Since these instruments, for gas, oil, and temperature, operate intermittently, the noise they create is neither constant nor in step with engine and road speed.

A technique for locating a noisy gauge is to listen while the noise occurs, then disconnect the "hot" wire at the rear of the gauge. This should be done to one instrument at a time, so that the troublesome unit can be pinpointed.

Treatment for the noise is accomplished by installing a 0.1-$\mu$F coaxial capacitor at the rear of the gauge, keeping the lead from gauge terminal to capacitor as short as possible. The metal mounting lip of the capacitor must make a tight, clean connection to the metal of the dashboard or other electrical ground.

# 23

## Accessory Noise

Electric motors which are used to operate various accessories in an automobile may occasionally give rise to CB interference. Usually the motors have carbon brushes which produce sparking that either radiates interference via the air or impresses it onto wiring which is ultimately connected to the transceiver. Among these accessories are the windshield-wiper motor, heater fan, and blower motor. Simply turning the accessory on and off while the transceiver operates will determine if it is causing noise.

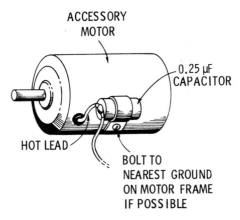

Fig. 23-1. Eliminating accessory-motor noise.

Minimizing the interference is done by providing a suitable bypass capacitor at the hot power lead that supplies the accessory. This is done, as illustrated in Fig. 23-1, with a high-quality, single-ended capacitor rated at 0.25 $\mu$F. It is mounted close to where the power lead emerges from the accessory. For maximum effectiveness, the body of the accessory should be well-grounded to the car frame with a short, heavy piece of shielded braid.

# 24

## RF Sniffer

After an automobile has been treated for noise reduction with the techniques described, there may remain noise sources which are difficult to pinpoint. One source is electrical wiring which runs close to such noise producers as the spark plugs, distributor, and generator. Lengths of wiring are impressed with noise energy which is retransmitted as from an antenna.

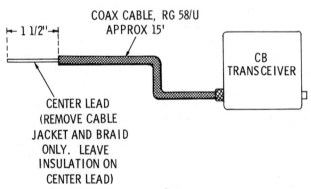

Fig. 24-1. Rf sniffer.

Determining which wires are the offenders can often be done with a simple device known as an "rf sniffer." It is nothing more than a length of coaxial cable, with a short piece of exposed center lead, connected to the transceiver. The arrangement is shown in Fig. 24-1. The unit is operated as a probe to locate noise sources. With the car engine and CB transceiver on, the tip of the probe is moved within an inch or two of various cables and wires. When strong noise energy is picked up, there is a corresponding increase in noise signal heard in the CB transceiver. These wires can be treated by bypassing with a coaxial cable capacitor, as described under Topic 21, "Ignition-Coil Noise."

# 25

## Wheel Static

Wheel static is a form of noise interference produced by the spinning action of a car's front wheels. Although not as common as other types of electrical interference generated by the automobile, it could prove helpful to know how to identify and correct it. In most cases it occurs when the car

**Fig. 25-1. Installing static-collector springs.**

is being driven at speeds over 20 mph. Static in the receiver is heard as popping sounds that change in character according to car speed. The noise does not vary according to the speed of the engine but with the speed over the road.

Wheel static can be identified by observing whether it drops out when the car is slowed down or stopped. To eliminate the engine as a possible source, permit the car to coast to a slow speed while the engine remains at idling speed.

Wheel static is caused by the build-up of static electricity between the front axle and front wheels. Between these parts is a coating of grease which acts as an electrical insulator. As the car rolls, the spinning wheel accumulates a charge in relation to the stationary axle. Then, as the wheel bounces, contact occurs between wheel and axle. This discharges stored electricity and produces static noise in the receiver.

A technique for eliminating wheel noise is with special devices known as "static-collector springs." By installing one at each front wheel, there is maintained a continuous contact, or electrical short-circuit, between axle and wheel. It prevents the build-up and discharge effect.

As shown in Fig. 25-1, the static-collector spring fits directly on the end of the wheel axle. This point is accessible after the decorative hub cap is pried off and the axle dust-cover removed. The spring is held in place, under tension, after the dust-cover is reinstalled.

Suitable springs are available from auto-supply outlets and some automobile dealers. Since the springs experience wear during operation, they should occasionally be checked (when the front-wheel bearings are repacked, for example) and replaced if necessary.

## Tire Static

In some cases the spinning tires of a car can build electrical charges which are heard in the speaker. Since it depends on motion of the car, this factor helps identify the source. The noise should change in pitch at varying road speeds. During these tests, the car ignition may be switched off to prevent its noise from masking that of the tires. Another symptom is that tire noise may sound differently when the road surface is wet. Moisture affects the discharge path of the noise voltage.

A special kit is available from many auto and radio part stores for eliminating tire noise. It consists of antistatic powder, actually powdered carbon,

**Fig. 26-1. Injecting antistatic powder.**

and a special tool (Fig. 26-1) for injecting the powder into the tire. In use, the tool is screwed on to the tire valve, then filled with a packet of powder. Then the tool opening is sealed with a cap, as shown. The fitting on this cap permits an air hose to force the powder inside the tire. The tool is removed after all four tires have been treated.

# 27

## Bonding

Unless adjoining sections of metal in a car are in good electrical contact, they can create or aggravate noise interference. Poor contact might occur due to corrosion, grease, or an intervening layer of paint. Bonding is a process of establishing electrical contact, either through cleaning or the addition of straps made from lengths of heavy copper braid. It not only helps reduce noise radiation, but it also reduces interference that may occur from the vibration of loose metal.

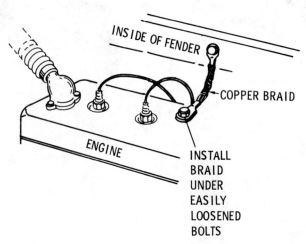

Fig. 27-1. **Bonding between engine and frame.**

An important point for bonding is between the engine block and car frame. Since rubber motor mounts can insulate the block from the frame, short pieces of braid should be installed between engine and frame at several points (Fig. 27-1.). The inside surface of the fenders, for example, are part of the frame. The car's muffler and tailpipe should also be grounded to the frame at the underside of the car. Leave enough slack in the braid to permit flexing as the car drives over the road.

Other points that may be bonded with braid: generator to engine block; mechanical linkages (from the carburetor, for example) to the block; and metal tubing or hoses. The hood may also be bonded to the frame. Provide sufficient slack for opening the hood.

Any point where there is a metal-to-metal contact, such as the mounting bracket of the ignition coil, voltage-regulator base, etc., should be cleaned with steel wool and the mounting bolts retightened.

# 28

## Appliance Filters

A number of appliances found in the home are prolific generators of radio noise and may interfere with CB reception. The source is usually simple to identify since the noise occurs only when the appliance is in operation. Some typical offenders include: vacuum cleaners, mixers, shavers, sewing machines, fans, office machinery, and air conditioners.

It is impractical in many cases to open the appliance and install bypass capacitors across a-c wires inside the appliance housing. Several plug-in filters, however, are available for the purpose. These units are inserted into the wall outlet and provide an outlet for the appliance plug. Choosing one of the major filter types depends on the degree of noise suppression required. For this reason, it may be helpful to try a filter on a return or exchange basis until you discover an effective type.

In light-duty applications, the simple one-capacitor filter may prove satisfactory. Next is the two-capacitor type, which generally has a ground lead. In severe cases of noise interference, the more complex coil-capacitor, or "all-wave," type might be indicated. Usually a heavier-current appliance generates more noise and requires the additional filtering provided by the coils. Be certain, however, that the filter's current rating—6 amperes in some cases—is the same or higher than that of the appliance it is filtering.

## Fluorescent Buzz

Due to their nature of operation, fluorescent lamps produce a buzz-type interference that may be picked up by a CB receiver. An important remedy is to locate the receiver at least four feet away from the fluorescent lamps. This reduces the pickup of noise energy radiated directly from the glass lamp. But, a portion of this energy may back up through the ac line and find its way into the receiver.

Locating fluorescent lamps as a source of noise is done simply by turning off the lamp and listening for a change in the receiver's noise level. Noise of this type sounds like a buzz since the lamp produces an electrical arc discharge in step with the alternating line voltage.

Some bypassing with 0.01-$\mu$F tubular capacitors rated at 600 volts can help reduce the amount of buzz entering the power line. During any of these steps be sure to remove ac power from the lamp by pulling out the plug of a desk lamp or removing the fuse in the case of a permanent fluorescent fixture mounted on a wall or ceiling.

Shown in Fig. 29-1A is a bypass capacitor installed in a fluorescent desk lamp that might be located near a CB transceiver. The lamp is opened and the point where the line cord enters the case is located. Strip away some insulation from each wire at this point and solder the capacitor across the

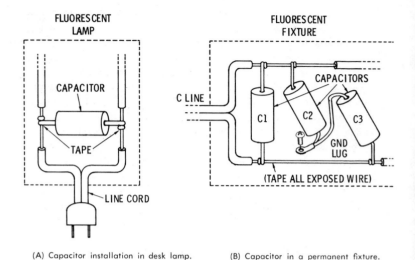

(A) Capacitor installation in desk lamp.    (B) Capacitor in a permanent fixture.

**Fig. 29-1. Eliminating fluorescent buzz.**

wires as shown. The final step is to wrap exposed joints with electrical tape to prevent short circuits. After the job is completed, turn on the lamp and CB set. By reversing the lamp's ac plug in the wall, it may be possible to obtain even further reduction in noise.

Fig. 29-1B illustrates the technique for suppressing buzz in a permanent fluorescent fixture. Here, three 0.01-$\mu$F, 600-volt capacitors are installed at the point where the ac power line enters the fixture. The first capacitor (C1) is installed exactly as already described for the desk lamp. The two remaining capacitors (C2 and C3) are each connected from the line to ground. The ground is the metal fixture itself, which, in most cases, is already connected to the electrical ground of the house wiring. A ground lug is very useful for connecting the two capacitor ground leads together and contacting them to the case. The ground lug is screwed to the fixture case after a small amount of paint (if present) is scraped from the case to ensure good electrical contact. Be sure to tape all exposed wires.

# TVI Stub

A simple filter can be installed on a television receiver to reduce a type of CB interference known as *fundamental blocking*. The filter, known as a *stub,* is fashioned from a length of ordinary twin lead, the flat 300-ohm ribbon normally used for the run between the television antenna and receiver. By adjusting the stub to correct length, it becomes a tuned circuit that effectively short-circuits the CB signal before it enters the television receiver.

The type of interference reduced by a filter of this kind is usually visible on more than one channel. In fact, it might occur throughout the television band. This is in contrast to the more common second-harmonic interference where a signal on 54 MHz (from the CB set) interferes mainly with channel 2. This is generally seen as a herringbone pattern on that channel. In fundamental blocking, the type to be treated here, there may be no such pattern, but a complete blocking of picture or sound.

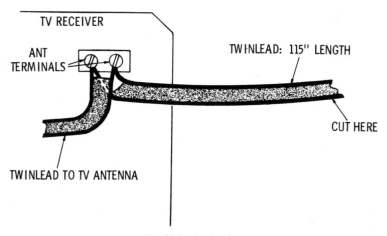

**Fig. 30-1. A tvi stub.**

The stub may be installed on your own television set, if desired, or on that of a neighbor's. It should be stressed, however, that in most cases of television interference (tvi), the CB operator is not obligated to install any filter on a neighbor's receiver. In more than 80 percent of all cases investigated by the FCC, the fault of tvi is usually that of the television receiver. For this reason, many television manufacturers offer suitable filters, to be installed by a serviceman, for correcting the problem. It should be mentioned, too, that if a CB operator installs any filter on a neighbor's receiver, there is considerable risk that he will be blamed for other problems which may develop in the receiver—although the filter is not actually at fault.

To install the stub, obtain a length of twin lead which is 115 inches long. Strip the insulation from the wires at one end and install the wires at the television set's antenna terminals. Note that the regular television antenna wire is left on the same terminals (Fig. 30-1).

Adjusting the stub to correct length is done while the interference is visible on the screen or heard in the speaker. Use a wire cutter to snip off about a half-inch of twin lead at a time while observing the interference. This is done at the end of the stub away from the antenna terminals. As the stub is shortened, it will tune to 27 MHz and considerably reduce its strength, resulting in less interference to the television receiver. The action of the stub should not interfere with normal television signals which commence at frequencies higher than 50 MHz. The completed stub should not be coiled, but stretched out behind the television set.

# 31

## TVI Filter

Where a CB set is located near a television receiver, there is risk of a type of interference which disturbs television reception on most channels. The filter illustrated in Fig. 31-1 is designed to reduce this interference. It is connected to the television antenna terminals. This circuit, known as a *series-tuned trap*, does not reduce another common type of interference, the kind which produces close-spaced lines across the picture. This is harmonic interference which must be reduced by a filter already inside the CB set, according to the manufacturer's instructions. The unit described here handles only the kind of interference which causes the picture to be completely blocked or torn.

The filter is constructed on any small piece of insulating plastic board with the dimensions shown. (Use experimenter's perforated phenolic board or bakelite, if you desire.)

Wind the coil on a dowel or rod of $1/4$ inch diameter, or even around a pencil. Then slip it off, remove the enamel insulation from the end wires, and mount the coil under the screws. The hardware may be three 6-32 machine screws and nuts. The capacitor is mounted by first flattening out its end tabs and placing the screws through the tab holes. The unit is completed by attaching a short piece of television twin lead under the two lower screws, then connecting the other end to the television set's antenna terminals. The set's regular twin lead from the antenna is left in place. When you are mounting the filter at the rear of the television, fasten it so its screws and metal parts do not touch anything. It can be held in place with tape.

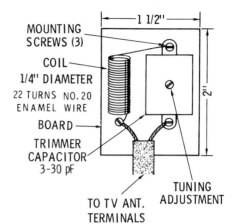

MOUNTING SCREWS (3)

COIL
1/4" DIAMETER
22 TURNS NO. 20
ENAMEL WIRE

**Fig. 31-1. A tvi filter.**

BOARD

TRIMMER
CAPACITOR
3-30 pF

1 1/2"

2"

TO TV ANT.
TERMINALS

TUNING
ADJUSTMENT

Tuning the trap is done while you are observing CB interference on the television screen. Adjust the capacitor tuning screw with a nonmetallic screwdriver until interference is reduced. In some cases, it may not be possible to tune out the interference, since the trap needs adjustment to its coil. First try spreading the coil turns slightly, then repeat the tuning adjustment at the capacitor. If this does not work, rewind the coil, adding about four more turns. Once the filter is correctly tuned, it should not interfere with normal television reception. The filter operates only on 27 megahertz, whereas the lowest television channel enters on slightly over 50 megahertz.

SECTION 3

# Equipment Maintenance

# 32

## Restoring Controls

Switches, potentiometers, and other controls accumulate a coating of oxide or air-borne grease that interferes with normal action after a period of time. The trouble may occur, for example, as a crackling noise in the speaker as the volume control is turned, or intermittent operation of a switch. Restoring the contact surfaces can be done with a spray can of contact cleaner, as shown in Fig. 32-1. A 23-channel crystal-selector switch is shown being cleaned, but the general procedure applies to other controls such as volume, squelch, slide, and toggle switches.

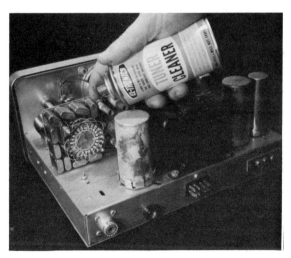

**Fig. 32-1. Cleaning controls with a spray cleaner.**

Be sure that the power is off and the line cord is removed from the wall outlet before beginning the job. Remove the chassis from its cabinet. Hold the spray-can nozzle near the control and apply a burst of cleaner directly on switch contacts or through openings in the control casing if the contacts are enclosed. It is especially important while spraying to keep the contacts in motion. In the case of a rotary switch, turn its knob back and forth; for a slide or toggle switch, snap it up and down. This ensures that the cleaner will reach all contact areas.

If the control is hidden behind other circuitry, a handy gadget to use is an extension tube usually supplied with the spray can.

# 33

## Cleaning Relays

After a period of time, the contacts on a send-receive (changeover) relay may become blackened with carbon, due to electrical arcing, or coated with foreign matter. The result is incomplete contact which interferes with circuit action. These contacts need to be cleaned, a job which should be done at intervals of approximately six months.

Certain precautions should be observed when working on a relay. Do not jostle the metal strips which hold the contacts; this could cause misalignment. Abrasive substances such as sandpaper or emery paper should not be used, since they might change the special shape of the contacts or cause short-circuiting.

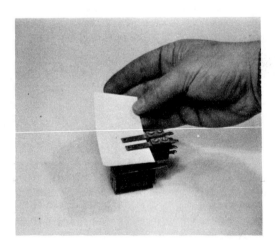

**Fig. 33-1. Cleaning relay contacts.**

In most cases, cleaning can be done with an ordinary business card or piece of heavy brown paper, as shown in Fig. 33-1. (The relay is shown removed from the chassis for clarity. Your relay can be cleaned while in place.) The card is inserted between the contacts and then it is drawn through them. As this is done, press down gently on the movable contact so that the card is squeezed slightly. Repeat this process several times, using a different area of the card for each pass-through. In many relays there are contacts above and below the movable center contact. For these relays, run the card both above and below the movable contact. This technique should suffice for most cases, but improved results are possible by first spraying the card with a cleaning solvent such as a tuner cleaner.

In older CB equipment, relay contacts may appear pitted from repeated sparking. Any attempt to remove pits with an abrasive file or paper could alter the shape of the contacts and introduce further problems. Most contacts are designed with a curve that causes a self-wiping action as they operate. Also, the shape is chosen for best electrical contact. To restore contacts that are in bad condition, use only the tool made especially for the purpose. This is an extremely fine file for "dressing" relay contacts. Such an instrument permits pits to be filed while removing nearly none of the essential contact surface.

# 34

## Electronic Lubricant

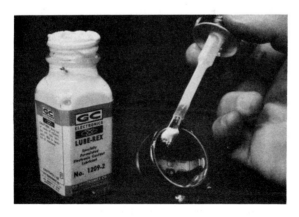

Fig. 34-1. A typical part requiring occasional lubrication.

Most of the mechanical parts in the CB transceiver undergo some deterioration with age. Spring tension relaxes, metal surfaces become worn, and dirt is deposited. An electronic-grade lubricant can revive these components and extend their useful life by reducing friction and improving contact action.

Typical parts which should be lubricated occasionally include the bearings of a variable tuning capacitor, mechanical linkages, and wirewound controls with exposed elements, such as the one shown in Fig. 34-1. Use the lubricant very sparingly or it will trap additional dirt. Also, move the part being lubricated as the lubricant is being applied to ensure that it reaches all necessary points.

# 35

## Scratchy Tuning

If a scratching sound is heard in the speaker while one is turning the channel dial of a tunable receiver, chances are that dirt is in the tuning capacitor. The cure is to blow out the spaces between the capacitor plates to remove dust. A soft brush may also be used. Then capacitor bearings should be lubricated with a silicon grease available for the purpose. Tighten the capacitor mounting screws. If the capacitor is mounted on rubber supports, don't crush them while tightening. If the rubber has hardened with age, it should be replaced. Otherwise it cannot provide shock-mounting which prevents chassis vibration from turning into audible sound in the speaker.

# 36

## Rotator Lube

Most antenna rotators that turn directional beams give years of trouble-free service. At approximately four- or five-year intervals, however, they generally require lubrication. Begin by eliminating any possibility of the rotator being operated while you are working on it; remove the control wires or the fuse that controls the power line. Open the housing which encloses the rotator motor and give it a thorough cleaning. Remove old grease and dirt as completely as possible.

The gears may then be lubricated with heavy grease. Do not use carbon-type or white grease. The grease may be the automotive type which can usually be obtained at a local service station. You'll just need a small quantity. After the gears are lubricated, the motor bearings may be lightly oiled to complete the job.

# 37

## Socket Leakage

A circulation of air under the chassis of a transceiver will, in time, cause a build-up of dirt or grease between the tube-socket pins. While this rarely causes short circuits, it can form an electrical leakage path that reduces the tube's operating efficiency. At approximately one-year intervals, the chassis should be removed and the bottom of each tube socket cleaned. This can be done with a toothbrush wetted with a radio-type solvent or cleaner. Be careful not to disturb the position of parts around the tube socket as you clean the spaces between the socket terminals. If there is any sign of dirt on the top of the socket, seen after the tube is removed, it should be cleaned with a cloth moistened with solvent.

# 38

## Grid Dipper

The Citizens band was conceived as a low-cost communications service. And one way to keep expenses low is to perform your own maintenance and servicing where possible. A licensed technician is required when troubles exist in the frequency-determining section of the transmitter. This is considered to be in the crystal-oscillator circuit(s). The receiver, and *final* radio-frequency amplifier of the transmitter may be adjusted or repaired without a license.

One of the most useful single instruments for troubleshooting and maintenance is the *grid dipper*, also known as the *grid-dip oscillator* (gdo). It is a miniature transmitter and receiver which is tunable to 27 MHz. It performs dozens of tasks that would normally require the facilities of a well-equipped test bench.

**Fig. 38-1. A tunnel dipper.**

When used in its oscillator function, the instrument is manually tuned to 27 MHz. This provides a signal equivalent to a carrier transmitted by a nearby CB transceiver. The advantage is that the signal is steady and available at any time. It is extremely useful in checking, troubleshooting, and aligning receiver circuits. It enables you to utilize service instructions provided either in the instruction manual or in available service literature.

When the dipper is switched to its "diode" function, it behaves like a miniature receiver on 27 MHz. This is convenient for transmitter checks. Modulation quality may also be monitored through earphones plugged into the dipper. The unit may also be used as a simple field-strength meter to check output and tuning adjustments.

The "dipper" function of the device refers to meter action when the unit is brought near a coil in a tuned circuit. By adjusting the dipper tuning dial and watching for a dip in the needle, you can determine if a tuned circuit is capable of tuning to 27 MHz.

The unit shown in Fig. 38-1 is actually a "tunnel dipper," a later version of the conventional gdo. The difference is that solid-state circuits in the tunnel dipper can operate from internal batteries. The regular dipper must plug into ac current. For this reason, the tunnel dipper might prove handier when making measurements at a mobile installation where house current is not conveniently available.

# 39

## Aligning With S-Meter

The i-f transformers in the receiver section of a CB transceiver rarely go out of alignment. But if a repair has been made in the intermediate-frequency section, or a new transformer installed, you may wish to recheck transformer adjustment. This is most quickly done by using the S-meter as a tuning indicator. A signal source is required; either a standard rf signal generator on 27 MHz or the steady signal from a distant station.

Permit the receiver to warm up for at least 30 minutes. With an alignment tool that specifically fits the slugs in the transformer, adjust top and bottom of each i-f can for maximum reading on the S-meter. Best results occur if you use as little signal as possible and the lowest readable indication on the S-meter. Repeat the adjustments several times since there might be some interaction between them.

# 40

## S-Meter Zero

After a period of use, the S-meter needle on some transceivers refuses to return to zero. Some units provide a control at the chassis rear for electrically zeroing the pin. This is described in the instruction manual.

In addition, the meter should be mechanically zeroed if a screw-slot appears near the bottom edge of the meter face. Turn the power off and use a thin screwdriver to adjust the meter pin to zero. As this is done, gently tap the meter face with a finger. This removes friction in the meter movement and permits accurate adjustment. If you notice that the pin refuses to return to zero, even with power off, the plastic meter cover may be charged by static electricity. This is removed quickly by gently wiping the plastic cover (from the outside) with a cloth moistened with water and a drop of liquid detergent.

## Squeaking Bearings

After a period of operation, mechanical rubbing in a transceiver's metal shafts may develop an annoying squeak. This can be silenced by applying graphite between the shaft and its supporting bushing. But powdered graphite, commonly available for the purpose, is not recommended. Since graphite is a good electrical conductor, spilled particles might cause short circuits.

The answer is a liquid-type graphite, which is actually powdered graphite suspended in a solution which evaporates after application. Widely available for lubricating locks, it is easily injected into a bearing with little danger to nearby circuits.

## Quick Battery Removal

Here's a tip borrowed from manufacturers of small transistor radios. It can be used effectively by owners of handie-talkies which use penlite-size batteries. Often, these batteries are difficult to remove from the recess formed by the holder. The cure is to place a strip of cloth, tape, or similar material in the bottom of the holder before battery installation. Allow the ends of the cloth to protrude out of the holder. When it's time for battery replacement, simply pull those ends and the cells come tumbling out with no fuss.

# 43

## Cable Check

A common cause of intermittent operation in a CB transceiver is a broken wire hidden inside a mike or antenna cable. The break frequently occurs near the connector. To locate the problem, operate the transceiver, and grasp the cable near where it enters the connector. Bend it gently in each direction. If normal operation is restored, you've located the break. This can also be done all along the length of a mike cable, which is usually subjected to much flexing and breakage after a period of time.

A similar problem occurs inside the connector where the wire may lose proper contact. While operating the transceiver, gently pull and push the cable where it enters the connector in an effort to check for normal operation.

# 44

## Locating a Defective Tube

One of two symptoms frequently pinpoint a dead tube. First is the absence of any red glow from the tube filament. Also, a dead tube is cold to the touch. A "click" test can also be used with some degree of success. It is done while the receiver is on and out of its cabinet. (Note that the following test is for the receiver section only and may not be used in the transmitter.)

Start from the audio output stage and work toward the front of the receiver. Remove the tube from its socket and listen at the speaker. As the tube is lifted, the circuit disturbance should produce a "click" sound. Return the tube to its socket. Continue with the remaining tubes, up to the receiver r-f amplifier. If the click is not heard while removing a given tube, this is an indication that the tube removed just *before* that one is defective.

# 45

## Longer Component Life

Probably the greatest strain on the components (including tubes and transistors) in a CB transceiver occurs when the power switch is turned on. There is a momentary surge that imposes somewhat greater loads on parts than when they operate in continuous fashion.

For this reason, you can extend component life by avoiding unnecessary on-off switching. Thus, if you intend to operate for several periods within, say, an hour, it would be advisable to leave power on during the complete hour rather than switching off during the inactive time.

# 46

## Fuse Holders

A small amount of dirt or corrosion inside a fuse holder can cause a voltage drop and some loss in efficiency. The effect is most noticeable when the unit is operating on mobile power, where supply voltage is normally only 12 volts. The fuse holder should be checked and occasionally cleaned.

With an in-line or cable-mounted fuse holder, it is easy to disassemble the holder and polish the contacts with a rough cloth. The job is slightly more difficult on a chassis-mounted fuse holder. (Be sure to remove the a-c plug from the wall outlet before servicing the holder.) Insert a blunt instrument about as thick as the fuse, and rotate it in order to grind away any coating on the inside contact.

# 47

## Tube Glows Blue

One cause of vacuum tube failure is the tube becoming gassy. It's due to a slow drawing of air into the glass bulb over a long period of time. It ultimately ruins the tube. In some instances, a slightly gassy tube will display a blue glow when observed in a darkened room. But extreme care must be used in judging this condition, since it is possible to discard a good tube by not interpreting the glow correctly.

Observe whether the bluish glow occurs *between* metal elements of the tube. This is easily seen by looking through the top of the tube and examining the space *between* the circular metal elements (Fig. 47-1). Glowing gas here is a sign that the tube should be checked. Gassy tubes draw excessive current in operation and could impose a strain on other circuit components.

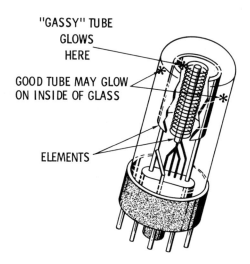

"GASSY" TUBE GLOWS HERE

GOOD TUBE MAY GLOW ON INSIDE OF GLASS

ELEMENTS

Fig. 47-1. Gassy tube.

Certain tubes, however, produce blue light during normal operation. (These are not to be confused with voltage regulator or mercury rectifier tubes which glow bright blue during operation. These types, however, are not normally used in Citizens band equipment.) Tubes which often glow are those in the output stages—audio- or radio-frequency—of the CB transceiver. Close observation will reveal that the glow is *not between* tube elements, but on the *inside surface* of the glass tube. Furthermore, the light is not uniformly spread through the tube, but shows up as flecks on the glass. According to one major tube manufacturer, such light is perfectly harmless and, in fact, is a sign of an exceptionally high vacuum.

# 48

## Special Generator

Two-way radio in many public-safety vehicles—police and fire, for example—requires high-charge generators or other heavy-duty power sources. Is this extra capability needed for a CB mobile transceiver? It is occasionally believed that a standard car generator or alternator may not be adequate.

This is a frequent misconception. No special charging or battery equipment is needed. Most police- or fire-department radios draw ten or more times the power of a CB set and need husky power sources. But the standard CB transceiver of the tube type consumes about the same or slightly more power than the regular car radio. This figure in many cases is about 50 watts (see manufacturer's specifications). Transistor equipment draws less than half as much as the tube types. The usual rating of an automobile generator or alternator is somewhat over 400 watts, and the standard electrical system of the car is more than adequate.

In some instances, there has been an attempt to increase power output of the transceiver by adjusting the car's voltage regulator. This increases voltage supplied to the set and, in fact, may raise the set's power output by a small factor. This negligible advantage, however, is cancelled by several problems created by higher-than-normal voltage. It can shorten the life of tubes and other components. If the CB rig is a transistor type, it might "blow" the semiconductors.

Overvoltage also extracts a toll from the automobile's electrical system, ending in an overcharged battery or a loss of melted solder from the generator armature.

If voltage in a 12-volt system is measured while the engine is operating at fast idling rpm, the reading should be slightly over 14 volts; for a 6-volt system it should be half as much. During slow idling speed (about 500 rpm) this voltage generally drops. For this reason, if transmitting is done while the car is parked, it is usually advantageous to race the motor slightly to obtain the higher, though normal, supply voltage, if the air has a generator. Alternator-equipped vehicles usually don't require this since they charge at low engine rpm.

## Extend Battery Life

If you operate a Handie-Talkie or other battery-powered transceiver and wish to conserve the batteries, here are methods for prolonging cell life. These techniques are for the dry-cell type of battery that is normally discarded after it wears out.

All dry cells discharge even when not actually in use. This is due to loss of moisture and a continuing chemical reaction within the cell. Both damaging factors can be greatly reduced by lowering the cell temperature. Tests have shown that the chemical activity in a dry cell (of the zinc-carbon type) comes to a halt at —22° F. Other measurements indicate that cells stored for two years at room temperature retain only 50 percent of their charge; while those stored at temperatures below zero can hold about a 90 percent charge.

This can be translated into practical terms for the user of battery-powered equipment. In many cases, a portable transceiver might be used only occasionally. Here it becomes feasible to remove batteries to prevent slow discharge when the unit is not in service. The batteries may be stored in a refrigerator, which usually maintains a temperature of 40° F.

Even greater improvement in shelf life is possible by storing cells in a home freezer. Temperatures in this instance are in the vicinity of 0° F. There is one precaution: condensation of moisture on the cells may crack the jackets and increase electrical leakage. For this reason, batteries should be placed in some protective coating such as plastic wrap. Also, the cells should be given time to reach room temperature before placing them in operation.

It is possible to recharge a worn cell a limited number of times. Chargers are available for the purpose. According to one large battery manufacturer, the cell should be charged very soon after removing it from service. Charging time is 12 to 16 hours and ampere-hours should be 120 to 180 percent of the ampere-hour discharge. Cells should be placed back into service soon after recharge, since their life on the shelf is very short.

# 50

## Printed-Circuit Crack

As CB equipment grows smaller, due to the use of transistors and other solid-state techniques, there is an increasing use of printed circuits. These circuits are reliable and rugged in service but may be damaged due to dropping or other mishandling. One symptom of damage is intermittent operation; the transceiver, for no apparent reason, performs properly at certain times and not properly at other times. A frequent cause is a small crack which has developed in the copper foil "wiring" etched on the plastic board.

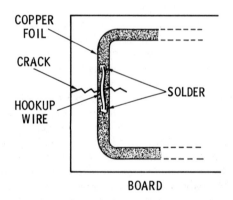

**Fig. 50-1. Repairing a cracked printed-circuit board.**

Since the crack is not often visible under normal room lighting, try to locate it by placing a strong light *behind* the circuit board. View the board from the other side. The crack is often revealed as a hairline.

Repairing the crack can be done by using the technique shown in Fig. 50-1. A "bridge" of wire is made to span the break. Use ordinary hookup wire (No. 20 tinned copper, for example). With as little heat as possible, solder the wire across the break, as shown. Excessive heat can cause the copper foil to separate from the plastic board.

# 51

## Transistor Damage

Transistors are known for their long life and low failure rate. However, to deliver these benefits, there are certain precautions which must be taken to protect semiconductors from accidental damage. Proper power-supply polarity, for example, is extremely important when installing a transistorized rig in an automobile. Be sure to check whether the car has a negative or positive ground, and make the installation accordingly.

Less well-known is the effect of mounting position within the car. Transistors are considerably more sensitive to heat than are vacuum tubes. For this reason, do not mount a solid-state transceiver in the direct flow of hot air from the car heater. It is possible that thermal protection provided by the manufacturer (such as heat sinks and thermistors) will not function properly at excessive heat levels. Some forethought will suggest the best mounting point for the transceiver.

Once the set is in place, determine whether any metal cables are touching the rear of the cabinet. If transistors are exposed at this point, there is some possibility of a short circuit developing if the transistor case comes in contact with grounded metal such as a mounting strap, choke cable, etc. Allow several inches of free space behind the cabinet to allow cooling air to circulate. Also, avoid operating the set immediately after entering a car on a hot day. If the windows have been closed, temperature may reach well over 100 degrees. Let the air circulate before turning on the transceiver.

Never operate a transistorized set without a proper load—an antenna or dummy load—connected to the output socket. It is easily possible to burn out the final radio-frequency transistor(s) when the mike button is depressed. This is due to a build-up of voltage in a tuned circuit connected to the transistor. The voltage may reach levels which puncture the semiconductor elements and cause complete failure. Such voltage is not developed when the circuit is properly loaded.

# 52

## Locating Microphonic Tubes

When the supporting structure inside a tube loosens with age, it allows the internal elements to vibrate and produce a ringing or howling sound in the speaker. The ringing is usually heard when the tube is jostled or disturbed, as in mobile operation. It can also be triggered by the normal movement of the speaker cone vibrating the chassis. To determine whether a tube is microphonic can be done by tapping on the transceiver cabinet with the knuckles. (The CB set should be on Receive, with the volume control at normal listening level.) If a ringing sound is heard while you are tapping, proceed to the next step.

Remove the chassis from the cabinet to expose all tubes. While listening carefully, tap each tube with end of a pencil, to isolate the one producing the ringing noise. More than one tube may seem to produce microphonic sounds. This occurs as the mechanical disturbance, applied to a good tube, couples through the chassis to the bad one. In this case, tap very gently and attempt to locate the tube which produces the loudest ringing in the speaker.

# 53

## Loose Chassis Grounds

At least once a year, it is advisable to retighten every accessible mounting nut and bolt in a CB cabinet and chassis. Some, you may find, have loosened from vibration or age. They can cause a variety of hard-to-find troubles in the circuit, from noisy performance to hum and intermittent operation. Retightening reduces electrical resistance that creeps into imperfect grounds. Don't overtighten hardware; it doesn't take much friction for a screw or nut to scrape away oxidation and establish a good contact surface.

# 54

## Replacing Tubes

After approximately one year of transceiver operation, it is recommended that vacuum tubes be tested on a reliable checker, preferably of the mutual-transconductance type. The test may reveal certain developing defects that could cause failure just when communications are needed most. Aging tubes develop partial short-circuits, reduced output, open filaments, or suffer from the entry of air (gassiness). These tubes should be replaced if such trouble is indicated on the checker.

A tube should be replaced with one of the identical number (with a few exceptions, as described below). Most efforts to substitute different tube types to achieve better performance are usually unsuccessful. They could lead to overheating and short life of components in the set's power supply or related circuits. Improper tube replacement may also burn out panel meters and possibly cause technical violations in equipment performance.

An exception is choosing an improved tube of the same number. The improved tube is usually labelled with a letter following the last character. For example, a 6AQ5 can be replaced by a 6AQ5A; or a 5U4G can be replaced by a 5U4GB. This works when letters are *added* but does not necessarily apply the other way around. An engineer, for example, might have designed a circuit to operate with a 5U4GB. It should not be replaced with the earlier 5U4G or 5U4GA.

Another exception to tube replacement occurs with special numbers produced by most tube makers. They are classified variously as "premium, high-reliability, special," and similar designations. Also, they are intended to directly replace standard numbers. The prime purpose of these tubes is to meet the needs of military and industrial users. They can withstand such factors as shock, vibration, high altitude, and heat. Some are tested with extreme care, others are guaranteed for about twice the life of an ordinary tube. For these reasons, the cost of such tubes is higher than normal. Installing these types in a CB transceiver will not produce dramatic differences in performance. In fact, the results may not even be noticeable. They will, however, provide the ultimate in tube reliability and performance. Whether the extra cost is justified must remain a matter of individual choice.

# 55

## Scope Monitor

One of the most accurate methods for checking the output signal of a CB transceiver is with an oscilloscope. The signal is introduced to the scope's vertical deflection plates. In Fig. 55-1 are several key patterns that may form on the screen. In Fig. 55-1A is the r-f carrier with no modulation. This occurs when the mike button is pressed, but without talking into the microphone. Strength of the carrier varies according to the pattern height.

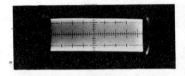

(A) R-f carrier, no modulation.

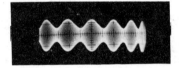

(B) Modulated carrier.

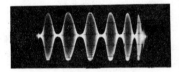

(C) Overmodulated carrier.

**Fig. 55-1. Checking transceiver output with a scope.**

The modulation envelope formed when one is speaking into the mike (or feeding in a steady audio tone for testing) is shown in Fig. 55-1B. 100-percent modulation, the maximum permissible amount, is seen by how audio pinches the carrier at the pattern's center line. The center line should just barely appear during this condition.

Overmodulation is seen in 55-1C. The widening of the center line indicates excessive audio or insufficient carrier. Such a pattern produces splatter and other distortion products on the air.

# 56

## Cabinet Ventilation

Holes, louvres, or screening on the cabinet of a CB transceiver are intended to provide ventilation for internal components. Size and layout of the openings usually encourage convection currents; air heated within the transceiver rises and exits through the top, while cooling air is drawn in via openings at or near the bottom. Thus it is important for cooling action, which extends life of the internal components, to remain unobstructed. The back of a transceiver should not be pushed against a wall, nor should its sides be covered by books or other surfaces. Try to avoid placing a transceiver just below a shelf. Several inches of free space should exist around the cabinet to encourage a continuous flow of cooling air.

# 57

## Connector Problem

One large manufacturer of CB antennas has noticed that a major cause of poor range can be traced to the connector that joins the coaxial cable to the bottom of the antenna. The problem tends to occur approximately six months after antenna installation. The reason for the trouble is that the standard coaxial connector (PL-259, for example) is not intended for all-weather operation. Rain and moisture can enter the connector joint even when the plug has been tightly screwed to the socket. The result is corrosion, which interferes with normal operation.

The cure is to leave no coaxial connector directly exposed to the elements. It can be protected in several ways. One is to spray the connector with a coating of acrylic or silicon rubber compound (after the connector is in place). Another method for waterproofing the joint is with a wrapping of plastic electrical tape lapped for several turns over the connector joint.

# SECTION 4

# Accessories

# 58

## Adding Earphones

With the addition of a few components, it is possible to modify many transceivers for earphone output. Earphones permit private listening and mute the speaker at the same time. A switch selects speaker or earphone output, and a jack allows headphones to be plugged in when desired.

Before starting, check to see whether the particular transceiver can be adapted for this circuit. Compare the schematic of the set with the one shown in Fig. 58-1A. The important requirement is that it must have one side of its speaker connected to chassis ground.

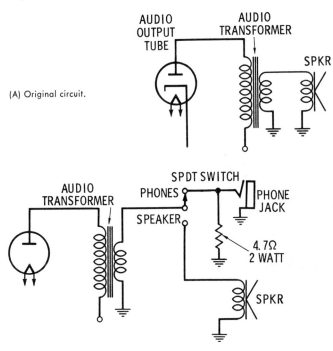

(A) Original circuit.

(B) Modified circuit.

**Fig. 58-1. Adding earphones.**

As seen in Fig. 58-1B, the speaker's "hot," or ungrounded, connection is broken and the leads are connected as shown. The additional components are: an spdt switch (toggle or slide type); a 4.7-ohm resistor, which acts as the proper load during earphone operation; and a standard phone jack.

These parts may be installed either within the transceiver cabinet or mounted in an external box. If a box is used, be sure to use a grounding wire between it and the transceiver chassis.

In operation, the switch is placed in the earphone or speaker position, depending on the desired operation. Virtually any type of magnetic earphone will operate when plugged into the phone jack, including the inexpensive permanent-magnet types in the 1000- to 3000-ohm category.

# 59

## Boom Mike/Headset

A boom mike, one that holds the mike in front of the mouth for hands-free operation, can be added to many CB transceivers. A model like the one shown in Fig. 59-1 combines a headset and boom mike into one accessory that's convenient for a busy operator engaged in continuous communications for long periods. Boom mikes are very popular among professional pilots because they facilitate talking without microphone grabbing and fouled cords.

Fig. 59-1. Combination headset and boom mike.

A boom mike may be supplied with a 2000-ohm ceramic element to match most CB inputs. Since connector styles vary, the cord is usually unterminated and the correct mike plug must be installed on it. A push-to-talk switch provided with the boom mike replaces the switch on the CB set regular microphone. To wire the headset circuit, follow the instructions already given in Topic 58, "Adding Earphones."

If you purchase a boom mike/headset, note that it is available with one or two ear pieces. The single-earpiece model is recommended if you intend to use the accessory in mobile operation. Motor vehicle laws in some states require that at least one ear be free while driving.

## Accurate Dummy Load

According to FCC law, a transmitter being operated during tests and troubleshooting should not be connected to an antenna except for extremely brief transmissions. For this reason, a dummy load to convert output power into heat is recommended. An ordinary No. 47 pilot lamp will work, but it provides only an approximate match into a transmitter with a 50-ohm output impedance. Since the lamp filament acts as a coil, it cannot perfectly substitute for a normal transmission line and antenna. An accurate dummy load, however, can be easily constructed from several resistors.

As shown in Fig. 60-1, the unit is made of three resistors wired in parallel. Each resistor is rated at 150 ohms, with a tolerance of five percent and a power rating of two watts. Note that they are wired and soldered together by their own leads. The total resistance of this combination is 50 ohms, and therefore it provides a good match to the transmitter. The two free leads of the dummy load are soldered to a short length of coax cable which is connected to the CB antenna socket.

The dummy load is capable of handing six watts of power, or about double the output of a typical CB transmitter. The load, however, will get warm in operation as it converts rf energy into heat. For this reason, allow a slight air space between adjoining resistors when wiring them together. This aids air circulation and, consequently, cooling. The only precaution is to keep all leads as short as possible to prevent a coil effect that could upset the 50-ohm impedance presented to the transmitter.

The shield can for the dummy load can be any convenient size. Note that one end of the load is bolted directly to the side of the can, while the

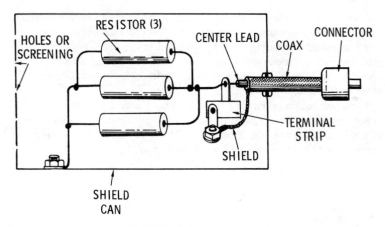

**Fig. 60-1. An accurate dummy load.**

other end goes to a terminal strip which keeps it insulated from ground. The center lead of the coax cable also conects to this point. Connect the coax shield to ground, which is the mounting foot of the terminal strip. A rubber grommet in the hole where the coax enters the can prevents the metal from cutting the cable.

The other end of the can should be sealed, except for several small holes drilled into it for cooling purposes. If you wish to use an ordinary tin can for the shield, this end can be where the top was removed. Some metal screening or scrap sheet metal should be used to recover the opening to prevent any escape of rf energy.

# 61

## RF Meter

A handy accessory for transceivers not equipped with an output indicator is a meter which indicates the presence of radio-frequency energy in the line running to the antenna. It is valuable for several functions. Each time the mike button is pressed, the meter pin reveals whether power is in the line. Thus it warns of low output when a fault develops in the transceiver. It is also a convenient tuning aid. Adjustments in the transceiver output stage can be tuned for highest reading on the meter.

The circuit illustrated in Fig. 61-1 can be constructed within a small metal box. It samples a negligible portion of the signal, and converts it to a form which can be indicated by the meter. In building this circuit, follow the component values shown and keep all leads as short as possible, especially the one which runs from the incoming cable (from the CB set) and the SO-239 connector to the antenna.

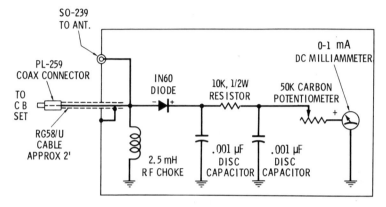

Fig. 61-1. R f meter circuit.

A recommended parts layout is to mount meter and carbon potentiometer on the front face of the metal box. The cable from the CB transceiver may enter from the left, and the SO-239 output connector on the right. Small components within the case are mounted on insulated terminal strips.

The potentiometer serves as a sensitivity control and prevents the meter pin from reading off-scale. It is adjusted by turning on the transmitter and turning the potentiometer until the meter pin reads about halfway up the scale. Each subsequent time the transmitter is operated, the pin should come to rest at about that point to indicate that everything is operating correctly. As you speak into the microphone the pin may flicker slightly under the effects of modulation.

## CB/AM Couplers

A special coupler is available to permit a CB antenna to operate for both CB transceiver and regular car radio. Instructions provided with this device are supplied by the manufacturer. Some CB operators have noticed a deterioration in reception on the car's regular broadcast radio after installing the coupler. In some cases this can be traced to poor antenna trimmer adjustment on the broadcast radio. This should be checked after the coupler has been installed.

The procedure is simple: find where the antenna lead emerges from the broadcast radio. Within an inch or so of this point there may be a hole that permits access to a slotted screw head. Some radios have the antenna trimmer located behind one of the control knobs. Access to the trimmer is obtained by removing the knob. Usually there are instructions on the radio case for correct adjustment, but the general procedure is to tune the radio to a weak station in the upper part of the broadcast band. The trimmer screw is then adjusted for maximum sound.

## Second Speaker

It is frequently convenient to hear incoming calls at some distance from the CB transceiver. The set may be located in a kitchen, for example, and calls missed when the operator is in the basement or attic. Extending the audio output of the transceiver can be done by the system shown in Fig. 63-1. This is a second speaker wired in parallel with the original speaker in the set. An external volume control permits adjusting loudness at the distant point.

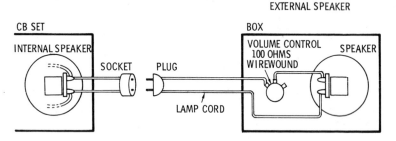

Fig. 63-1. Adding a second speaker.

The audio signal is picked up directly at the terminals of the internal speaker. A pair of wires is brought out of the set and fitted with a socket. This allows the external speaker line to be rapidly disconnected at a future time. The line is run to the distant speaker with ordinary lamp cord (No. 18 wire). Distances up to about 50 feet should cause little loss in sound volume. There will, however, be a slight drop in sound level at the transceiver's own speaker. But this can usually be compensated by simply running the regular volume at slightly higher level. The external speaker should be similar in type to the one already in the transceiver and have the same impedance, usually 3.2 ohms.

In this simple hookup, both speakers produce sound simultaneously. Also, the maximum volume level at the distant speaker is determined by the regular volume control at the transceiver. The control on the external speaker can only decrease the loudest sound provided by the regular control.

# 64

## Installing RF Indicator

Many older CB transceivers have no rf indicator on the front panel to reveal whether the transmitter is developing ample output. The indicator illuminates when the mike button is depressed and flickers with modulation. A neon bulb can be installed inexpensively to provide this convenience. It works with tube-type, 5-watt transceivers (not transistor types). Also, the final rf output tube of the transmitter should be located within a few inches of the front panel.

The neon lamp, an NE-2 type, is mounted in a small hole cut in the front panel, as close as possible to the tube socket of the final rf amplifier stage. One method for retaining the lamp to the panel is with a small rubber grommet inserted in the panel hole, with the neon lamp inserted into the grommet for a friction fit. One lead of the lamp is connected with a piece of hookup wire to the plate pin of the rf amplifier tube socket. Don't permit this lead to run close to chassis metal, and keep it short. Cover any bare wires with spaghetti insulation.

The remaining lamp lead connects to a 1-meghom $\frac{1}{2}$-watt resistor. The other resistor lead connects to chassis ground. After the job is complete, operate the transmitter. If the lamp remains on continuously, even during receive, increase the value of the 1-megohm resistor until the desired action occurs—illumination only when the carrier is on. Since the lamp requires only about 40 milliwatts during operation, it consumes negligible power from the transmitted signal.

# 65

## Spare-Fuse Holder

Small spare items, like a fuse, have a way of disappearing when they're needed most. You can take a cue from military electronic equipment and install a holder expressly for a replacement fuse. Parts distributors often stock a clip-in type holder to fit any standard fuse size. These holders usually have a black plastic base that can be mounted either on the chassis inside the transceiver, or on the rear panel if space is available.

# 66

## Paste-On Guide

A transceiver's instruction manual is often the best source of important information such as required tube and fuse types. But it may be misplaced or not available when you are engaged in mobile operation far from the base station. It takes just a few minutes to prepare a small card containing important data in condensed form and to paste it on the rear or bottom of the cabinet for future reference. It should contain a simple drawing which shows the layout and number of each tube used in the rig. The fuse type and its value should also appear. Other items, such as position of receive and transmit crystals might also prove valuable at a later time.

# 67

## Home-Brew Cards

If you exchange QSL cards you probably agree that the most distinctive and appealing ones are those with a personal touch. You can home-brew cards that contain such special features as a photo of yourself, the equipment, a cartoon, etc., without too much trouble or expense.

Prepare a card containing all necessary information and illustrations. Mount photos, or draw with India ink. For neat lettering that looks printed, try one of the currently available "dry transfer" sheets. The completed card is then photographed to obtain a negative. Unless you have some skill as a photographer and know how to make a copy negative, the job might best be left to a professional.

Once a negative is secured, any number of prints, which become the QSL cards, can be duplicated. Use double-weight print paper trimmed to postcard size.

# 68

## Earphone Pads

Earphones cause discomfort after being used for extended listening periods. Pads for easing the pinch of phones are often available. But you can also fashion a pair for yourself by cutting them from foam rubber and cementing to the phones. The rubber is available at well-stocked hardware stores. If your phones don't have an adjustable headband, you might have to bend the band so the phones won't exert excessive pressure against the ears.

# 69

## Crystal Bank

A handy way of storing crystals is in a holder, like the one in Fig. 69-1. It permits quick, easy channel selection. Crystals are plugged into the holder in numerical order. Use heavy cardboard lengths and fold over the sides, as shown. Holes for the crystal pins are made by a sharp-pointed instrument.

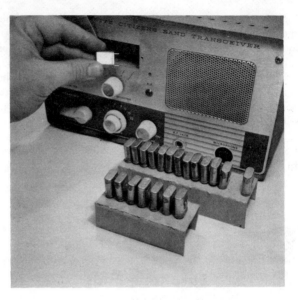

Fig. 69-1. Crystal holder.

A more elaborate holder may be made from a strip of wood, drilled with suitable holes for receiving the crystal pins.

# 70

## Car Plug-In for Handie-Talkie

Many Handie-Talkies rated at a power of 1 watt or higher are generally equipped with 12-volt batteries, or they may be powered directly from a car's electrical system. In the smaller 100-milliwatt category the hand-held transceiver is frequently powered by a 9-volt battery. This makes it impractical to eliminate the battery and run the unit directly from the car. Such usage would apply excessive voltage to the transceiver circuits and possibly burn out transistors.

The device described here is designed to overcome this limitation and make it practical to operate the small 9-volt transceiver with a plug inserted into the car's cigarette lighter or through a wire attached to a 12-volt power lead (the one running to the regular car radio, for example). The device is easily and inexpensively constructed using the diagram illustrated in Fig. 70-1. It is based on a zener diode in a voltage-regulator circuit.

The circuit fulfills several requirements. It drops car voltage from 12 volts to 9 volts to power the transceiver. It smooths out variations in voltage to the set; the car's electrical system can swing from about 12 to 16 volts during normal operation. Finally, the device overcomes the shifting electrical current drawn by the transceiver.

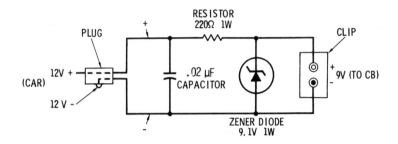

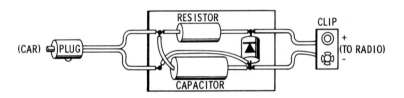

**Fig. 70-1. Car plug-in for a Handie-Talkie.**

Here are the required parts needed for construction:

*Plug*—This is a plug designed to fit into the car's cigarette lighter socket. It is available from either automotive or radio supply outlets. These plugs have two wires emerging from a plastic body. You must identify which wire runs to the tip contact of the plug. This is done by measuring with an ohmmeter, or simply by hooking a battery and lamp or buzzer to check which wire runs to the tip. Once it is found, mark it as the positive lead. The other wire, the negative connection, is at the side of the plug. Proper polarity must be observed or the zener diode and radio will be affected.

*Capacitor*—This component can be any capacitor—paper, tubular or ceramic—rated at 0.02 $\mu$F and with a voltage rating of at least 100. Purpose of this component is to help filter out noise from the car's electrical system which might reach the radio. If the capacitor has a dark ring or band at one end, connect this end to the negative side of the circuit.

*Zener diode*—This is the solid-state regulator. Its specifications are 9.1 volts at 1 watt. A typical unit is the GEZD-9.1. The important consideration is hooking it into the circuit properly. If you closely examine the diode leads, you will note that one lead is insulated from the metal case. This lead connects to the negative side of the circuit. If there is a symbol imprinted on the diode, the negative side is the arrowhead side; the bar marking goes to the positive side.

*Resistor*—This is a 220-ohm resistor, rated at 1 watt.

*Clip*—The output voltage of the circuit (9 volts) is most conveniently introduced into the Handie-Talkie by a matching clip that attaches to the battery clip already in the radio. This can be done by removing the clip end from an old 9-volt battery and soldering it to the + and − wires from the device. Carefully check to see that polarity is correct; that the + side of circuit does, in fact, connect to the + battery connection in the transceiver. Use the original 9-volt battery as a reference. It should indicate which side of the clip is +.

The complete circuit is just a few inches square and can be conveniently built into a small plastic or other type of nonmetallic box.

# 71

## Output Selector

This is an accessory which enables you to switch instantly between various antennas or other outputs used in CB trouble-shooting. It eliminates the bother of unscrewing the coaxial connector at the rear of the transceiver and hooking to a different antenna or, for example, a dummy load. Constructed in a small aluminum box, it is positioned next to the CB transceiver, or anywhere within arm's length.

As shown in Fig. 71-1, the input socket goes to the CB transceiver. The selector is a five-position switch (Mallory 3115J) that connects the signal to any of five output sockets. Connected to the output sockets are antennas, dummy load, test equipment, etc. It is not necessary to utilize all sockets, and several may be left open for use at a future time. One possibility is to connect a No. 47 pilot lamp to one output socket. Switching to this position provides a quick check on relative power output. The lamp will glow brightly with modulation.

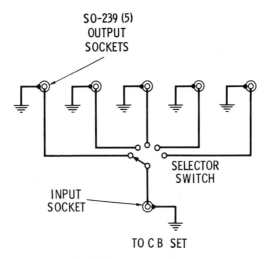

Assembly of the unit can be done by mounting the six SO-239 coaxial connectors on a small metal "minibox." A convenient arrangement is to mount the switch and input socket on one side, and five output connectors on the other. Keep all wires in the box very short and free of curves. Lengths of No. 18 solid, tinned copper wire are recommended. When mounting each socket, be sure its body makes good contact with the metal case. Scrape away any paint that might prevent a low-resistance electrical connection.

There is one precaution in operating this device, especially if you have a solid-state transceiver. Do not switch from one position to another while the transmitter is in operation. This might damage the transistors since there is no proper load between switch positions.

# 72

## Audio Compression

This add-on accessory might be considered for those transceivers which do not already have some built-in method for keeping average modulation as close to 100 percent as possible. In providing this function, the device can markedly improve operating range, or more precisely, intelligibility of the voice at a distant receiver. The audio compressor is similar to avc, or automatic volume control. If voice tones are soft during certain syllables, they are automatically strengthened. The reverse action occurs for excessively loud tones. A compressor is generally inserted between the microphone and its panel socket on the transceiver with no internal connections required.

# 73

## Improved Sensitivity

Transceivers that suffer from poor sensitivity tend to miss weak incoming signals. The circuit is unable to amplify low-level signals effectively without boosting interfering noise at the same time. Such circuits are also susceptible to a type of interference known as the "image." This is a false incoming frequency which should be rejected or weakened.

A cure for poor sensitivity is an external booster amplifier especially designed for CB frequencies. It is variously known as an r.f preamplifier or preselector. Either of these accessories can significantly pep up performance in a low-sensitivity receiver.

# 74

## Better Selectivity

CB receivers that are unable to separate closely spaced signals are said to have poor selectivity. It results in interference, for example, on Channel 11 when a strong station operates on Channel 10 or 12, the two adjacent channels. Circuits with good selectivity usually include some kind of filtering (mechanical or crystal) or a circuit known as "dual-conversion."

Sets with none of these refinements may be improved with an accessory known as the "Q-multiplier." The device, which connects into the CB circuits, effectively narrows receiver bandwidth. Usually there are two control knobs—*peak* and *notch*—which enable the operator to reject much adjacent-channel interference. The Q-multiplier is available either as a kit or in wired form.

# 75

## Line-Voltage Problem

The life of vacuum tubes in a transceiver is affected by line voltage; not only when voltage is excessive, but when it falls below the nominal 117-volt value. A clue to abnormal line voltage in the home is brightness of lamps or a television picture that shrinks and fails to fill the screen. A voltmeter gives a more precise indication.

A number of voltage-regulating accessories are available from electronics parts stores to help minimize the problem. Some are automatic regulators, others have a voltage-selector switch. A suitable unit might repay its cost in terms of extended tube life where line voltage isn't constant or correct.

*SECTION 5*

# Operating Techniques and Aids

# 76

## Evaluating S-Units

Most CB transceivers are equipped with an S-meter to provide an indication of incoming signal strength. A common misconception in reading such meters is that a doubling of the number of S-units, say from S4 to S8, refers to a doubling in signal power to the receiver. In actual operation the S-meter needle would barely move for a doubling in power. Depending on the particular receiver, signal doubling may appear as less than one S-unit increase.

Virtually all S-meters on CB receivers provide only an approximate reference to the incoming signal. Also, the S-meter readings between sets of different make, even when indicating the same S-reading, are not necessarily equal to each other. The reason is that a highly accurate meter circuit alone, with a standard calibration, may cost several times the price of a complete CB transceiver.

Here are suggestions on how to judge S-meter readings. Your ear, of course, detects whether an incoming signal is readable or not, but consider these points if the S-meter is used for checking equipment performance. When the S-meter needle varies in the lower part of the meter scale, from about S1 to S5 these changes represent small increases in signal strength. When the needle is in the upper part of the scale, the same amount of needle travel represents a far larger power increase. In other words, the higher the needle, the more power is required to swing it a given distance.

The practical application of this might be when one is attempting to check on differences caused by changing the position of a roof antenna. We'll assume that the original position produces an S4 reading on the meter when a known signal is received. After repositioning the antenna, the meter rises to S5 for the same signal. You may believe that an increase of one S-unit has provided a worthwhile gain in received signal strength. Now repeat the test beginning with a signal of S7, with the antenna in its original mounting position. When the antenna is moved to its new location, the pin may now rise only a fraction of one S-unit although the power increase is *exactly the same* as before.

For these reasons, checks of this type should be run with low S-meter readings, where any improvement will be easier to view. Also, there is a tendency of some CB transceivers to produce an increasing error on their S-meters as the signal reaches approximately S9 and higher.

The following chart reveals the basic idea. It was compiled by taking a reasonably high-quality CB transceiver and feeding into it increasing signal levels from an extremely accurate laboratory signal generator. S-meter readings are shown in the left column, while the signal levels which produced them are shown in the right column. As you can see, it took a disproportionate number of microvolts to cause the meter to rise from one S-unit to the next. (The chart begins at S3 since noise level in the receiver produced continuous readings below that figure.)

| S-Meter | Signal Input (in microvolts) |
|---------|------------------------------|
| S3 | 2.5 |
| S4 | 3.5 |
| S5 | 6.0 |
| S6 | 12.5 |
| S7 | 26.7 |
| S8 | 72 |
| S9 | 180 |
| + 10 dB | 400 |
| + 20 dB | 1200 |

# 77

## Mike Technique

Most transceivers have built-in circuits to prevent modulation from exceeding 100 percent, the legal maximum. Attempting to speak too loudly into the mike may result in distortion and difficulty in others' understanding your voice at another station. Since some mikes require "close-in" talking while others are sensitive at several inches, try to determine the best distance for your particular mike. A valuable aid for checking is a tape recorder. Have a friend record your signal off the air while you speak at varying distances from the mike. Also, try different loudness levels of the voice.

Recording the signal from a CB transceiver can be done by clipping a pair of leads across the loudspeaker and running them to the "Phono" or auxiliary input on the tape recorder. Try reversing the clips to the speaker to get the least hum on the tape.

# 78

## Handset Operation

In some applications, a CB set with a "radiotelephone configuration" (Fig. 78-1) offers certain advantages. Since it works like an ordinary telephone handset, you hold the receiver part directly over your ear and thereby eliminate noisy road conditions, the roar of a boat engine, or other interfering sounds. Also, you can hear weaker signals because the receiver part is equivalent to an earphone. The handset has a push-to-talk bar on its midsection to let you control the send-receive function while the instrument is held to the mouth and ear. When not in use, the handset is placed in a cradle on the transceiver. You can hear incoming calls from this position because the transceiver automatically switches to speaker operation whenever the handset is on the cradle. The speaker is cut off as the handset is lifted and audio is sent through the hand-held receiver.

# 79

## Operating Position

If your base-station rig is normally on a table top, operating convenience can be improved by preparing the easily constructed shelf shown in Fig. 79-1. Controls are easier to operate when they are raised from the table surface by a few inches. This arrangement also leaves a compartment below the rig for storing paper, crystals, and other items. Another feature is that the shelf can be constructed to tilt the transceiver at an angle. It makes dials and S-meter easier to view.

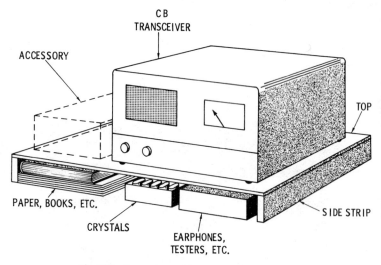

Fig. 79-1. Convenient table-top operating position.

Dimensions vary according to the rig, but you can experimentally determine the best height for the shelf. Try approximately three inches as a starting point. The shelf can be made of wood or hardboard, with wood strips serving as side supports. When making the side strips, be sure to cut them wider in front so the top edge of the transceiver tilts away from the eye. The shelf surface can be made wider than the rig to accommodate accessories, such as a speech compressor, if desired.

# 80

## Station Identification

A sizeable number of citations have been issued by the FCC for improper station identification. Let's consider the main requirements of this section of the rules. First is that call signs must be complete; each digit and number is to be spoken. Dropping the letter prefix and just giving numbers, e.g., "This is 4480," is not permissible. You may transmit in a language other than English, but in all cases call signs must be spoken in English.

There is no prohibition on adding numbers or letters to the call sign for personal or business use. These are termed "unit designators" and refer to the usual practice of saying, for example, "KBA4480 *Base* calling *Mobile*." Another example: "KBA4480 *Unit 1* calling *Unit 2*." In an exchange of this type, between two units of the same call sign, it is necessary for only one of the units to give the call sign. Although two units are in operation, they are considered under one station or call sign. It is permissible to eliminate such words as "This is" during station identification.

Consider, next, how often station identification must be given. According to the regulations, it is required before and after each transmission. However, if there is a *series* of transmissions or exchanges between stations, call signs need be given only at the beginning or end of the overall transmission period. This permits rapid back and forth communication. The overriding rule, however, is that when such exchanges are long, call signs must be given at least every 15 minutes.

Let's assume that communications are difficult and stations in contact decide to switch to a different channel. In this instance, all the rules of station identification apply as if communications were just beginning.

A minor exception to the rules is when one is establishing communications with a station whose call sign is not known. A name or trade name may be used for calling purposes, but after contact is made, the call sign of that station is to be used. Another exception is when a CB station is operated in signaling or remote-control fashion. No call sign is normally required. This would apply in the case of selective call, where tone frequencies may be transmitted without speaking the station call sign.

# 81

## Silent Period

Since the Citizens band is operated on a "party line" basis—many stations sharing a number of frequencies—transmission time is limited. The general rule is that transmissions be held to the minimum practicable time. Only during communications which concern the immediate safety of life or property does this rule not apply, and emergency transmissions may be continuous or uninterrupted. In such instances, however, the FCC requires that it be notified of such operation.

The specific time limit on transmissions is five minutes. This means five consecutive minutes of transmitting time between one or more stations. At the end of the communications (which may last less than five minutes) stations must remain silent for a five-minute period. This gives other operators an opportunity to use the channel. During the silent period, stations should monitor the channel to check for traffic before resuming transmission.

Consider some modifications and exceptions to the basic silent period. One is that when stations in contact with each other switch to a different channel, the total talking time is still five consecutive minutes. (This is written into the law to prevent evading the five-minute rule.)

One exception to the silent period is when you are called by another station who may not be aware that you are observing a silent period. In this case it is permissible to answer the caller and briefly acknowledge his call. You may inform him that you cannot communicate at this time, and for him to stand by until your silent period is over.

The silent period, as written in the regulations, applies to "communications between and among Class-D stations." For this reason, it is not interpreted to apply to units which operate under one call sign when in communications with each other. Although they comprise more than one unit, they are considered one station.

# 82

## Using Surplus Equipment

By purchasing CB equipment from a reputable manufacturer you are assured that it meets technical standards specified by the FCC. In fact, the literature of the manufacturer will certify that the circuits comply with existing regulations. This also applies to kit-type transceivers, where the manufacturer usually factory-assembles critical portions of the unit. In addition to conventional CB equipment, there are a number of military units on the surplus market, or in the hands of some experimenters, for which CB operation is claimed.

It is risky to use such equipment on CB since in nearly every case it will fail to comply with regulations in one or more of the following respects: It is incapable of maintaining frequency stability within 0.005 percent; much equipment in the surplus class utilizes frequency modulation (f-m), which is not permitted in CB; power input is in many cases beyond the 5-watt input maximum; and also, the surplus unit is often incapable of keeping within an 8-kHz bandwidth, as specified by law.

There are other considerations, such as harmonic output and proper trapping of signals which cause television interference on 52 MHz. Furthermore, much surplus equipment is designed for operation on unconventional power sources—such as 24 volts—which means that a new power-supply section would have to be constructed. A great deal of technical know-how, too, is required to convert equipment to meet CB regulations. Furthermore, the finished conversion would have to be checked and approved by the holder of a commercial radiotelephone license. Thus military surplus, attractive to the amateur radio hobbyist, is considered impractical for CB operation.

# 83

## Transmitting Precaution

Where road construction is in progress, the mobile CB'er may encounter a sign with the approximate wording: "Danger—Blasting, Turn Off Two-

Way Radios." This applies to CB equipment, which should not be used on transmit while in the vicinity of the warning. There is risk of detonating dynamite used in road-blasting operations.

Actually, the dynamite is normally exploded by an electrical blasting cap which is triggered through wires. There is a possibility, though remote, that a close CB signal may induce a flow current in those wires and set off the cap. A safe distance is generally considered to be more than 100 feet between antenna and the blasting cap. Allow a wide margin above this figure since your car may be passing over a roadway bridge, with the blasting operation directly below it.

# 84

## AC Outlet Ground

Provided on the back of most CB transceivers is a ground terminal. It is intended to connect to an electrical ground, the car frame in the case of a mobile installation, or a suitable ground in the home for base-station operation. Consider how to obtain the ground connection in the home.

**Fig. 83-1. An ac outlet ground.**

The CB instruction manual may recommend the use of a radiator or cold-water pipe for the ground connection. These are useful points when available. They are not, however, easily found in some rooms; and running an exceptionally long wire is not recommended. Overly long connections might reduce the purpose of the ground, which is to minimize noise pickup in the receiver, improve lightning protection, and possibly improve signal strength.

A particularly handy ground is available at the wall outlet where the transceiver is plugged in. If the outlet was originally installed according to the National Electrical Code, which applies in most towns, the screw which holds the cover plate provides a direct connection into the home's electrical grounding system. The screw makes contact with the metal outlet box which, in turn, runs to the building ground, either through armored cable or a ground wire. These ground connections then go to the building's water supply piping or a copper rod driven into the ground.

Hook a ground wire to the cover screw as shown in Fig. 84-1. Use a substantial conductor—No. 14 solid copper wire, for example. (It is not necessary for this lead to have an insulating jacket.) The thickness of the wire and the fact that the cover screw has a tapered head make a simple wrap-around connection unsatisfactory. Tightening the screw causes the wire loop to squeeze out. This can be prevented, as shown, with a solder terminal attached to the end of the ground wire.

When you are removing the cover screw there is a shock hazard from electrical terminals inside the outlet box. Thus it is recommended that the fuse for that outlet be removed before the ground wire is fastened in place.

# 85

## Upgrade a Crystal Mike

Some older transceivers were equipped with crystal microphones. Such mikes are poor performers under the humid or extremely hot conditions, that may be encountered in mobile operation. Loss of output and deterioration of signal can occur.

Crystal units are readily interchangeable by mikes of the ceramic type. Ceramic elements are highly resistant to heat and humidity and are electrically equivalent to the crystal mike. No internal circuit modifications to the CB set are needed.

# 86

## 6-Volt Set, 12-Volt System

In some instances, it is desired to operate an older CB transceiver, containing a 6-volt power supply, in a newer automobile with a 12-volt ignition system. It is possible to obtain a dropping resistor that lowers the 12 volts to 6 volts. Such units are available from radio supply sources. Obtaining a resistor with the correct rating is important; otherwise it is easily possible to burn out the resistor or provide incorrect voltage to the CB rig.

Determining the correct value for the resistor is done by consulting the CB instruction manual to find the amount of current, in amperes, consumed by the set when it is operated on 6 volts dc. This figure is divided into 6 to give the required resistance in ohms. For example, if the set requires 5 amperes, a 1.2-ohm resistor would serve.

Equally important is the resistor power rating. This is found by multiplying the CB set's current by the operating voltage. Therefore, in our example, it would be $5 \times 6$, or 30 watts. To prevent the resistor from burning out prematurely, the wattage rating should be approximately doubled to about 60 watts.

Exact resistance values, such as 1.2 ohms, may not be easily available. For this reason, resistors sold for voltage dropping are usually equipped with a slider that permits selecting the proper resistance for a particular rig.

# 87

## CB in Hi-Fi

Most cases of CB interference affect television receivers, but there are frequent instances where the CB signal enters telephone lines and hi-fi systems. Curing telephone interference is best left to the phone company which, in some cases, will install special bypass capacitors inside the instrument.

In cases of interference to hi-fi systems, the CB operator's voice is heard in the speaker. There are several measures that can help reduce this interference. The hi-fi system should be well grounded. This can be done by running a heavy wire from a chassis screw to the screw that holds the cover plate of the ac wall outlet. Next, install a bypass capacitor at the speaker leads. This can be a 0.01-$\mu$F capacitor rated at 600 volts. It is connected across the speaker terminals, either at the amplifier or at the speaker.

If interference persists, the hi-fi amplifier will have to be removed from its cabinet, and additional bypassing done under the chassis. Find where the ac line enters the chassis. From each leg of the line, inside the chassis, connect a 0.01-$\mu$F, 600-volt ceramic capacitor to ground. If the interference occurs only when the hi-fi system is operating on phono—but not on tuner or tape—a small bypass capacitor is recommended at the first audio amplifier tube, sometimes called the *phono preamp* stage. The grid of the tube is located, and a 50-pF mica capacitor soldered from grid to ground. Leads on this capacitor should be kept as short as possible.

Another measure which may reduce interference when one is playing records is to reduce the lengths of the cable running between the phono cartridge and input socket on the amplifier. It is possible for the cable, at certain lengths, to act like a resonant antenna for CB signals. Shortening the cable can upset this undesirable dimension.

# Tuning a Pi Network

A popular circuit for tuning a transmitter into the coaxial transmission line is the *pi network*. As shown in Fig. 88-1 there are two means of adjustment; each is a variable capacitor. The one nearest the final rf tube of the transmitter is the *plate capacitor*. The other is the *antenna capacitor*. (These may be marked on the transceiver or located by comparing this schematic with the diagram of your CB unit.) The capacitors require a careful tuneup the first time the transceiver is installed, and also each time it is connected to a new antenna. Two tuneup techniques are possible.

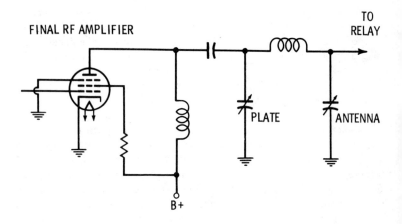

**Fig. 87-1. Tuning a pi network.**

The first method requires a milliammeter plugged into the jack found on many CB sets for checking rf power input. The antenna and coax line are connected to the transceiver, and the set is placed on transmit position. The first step is to adjust the plate capacitor for a dip on the meter (least possible current). Next, the antenna capacitor is adjusted for a very *slight* increase in current. (Under no circumstances should the plate capacitor ever be used to increase the meter reading.) Return now to the plate capacitor, and tune again for a dip. Continue this back-and-forth tuning until adjusting the antenna capacitor no longer increases the current. At this point the transmission cable will be absorbing maximum power from the transmitter. The last step should always be tuning the plate capacitor for a dip.

The final reading on the meter should agree with the manufacturer's rated current for producing full 5-watt input power, usually somewhere between 20 and 30 milliamperes. A reading in excess of this rating can be lowered by backing-off on the antenna capacitor.

The second tuneup method is done with a field-strength meter, which acts as a miniature receiver, placed at some distance from the transmitting antenna. Both plate and antenna capacitors are now tuned for the *highest* reading on the field-strength meter. The procedure, however, is done in careful steps, as before. Begin by tuning the plate capacitor for maximum output. Again a back-and-forth action is done with the adjustments to attain highest output. If the antenna capacitor at some adjustment point causes the meter reading to drop suddenly, turn it slightly in the opposite direction to restore transceiver output.

# 89

## Proper Use of a Field-Strength Meter

Most simple field strength meters tend to give exaggerated readings. Even small changes in transmitter tuning can cause the meter needle to swing way up and give the false impression of high output power. The first precaution in using most meters, therefore, is to consider all readings as simply relative. Numbers which appear on the meter face should not be taken literally. When the needle moves from 2 to 4, for example, it does not indicate a doubling of power, but a much smaller increase. The needle is primarily useful for its relative up or down movement.

Before readings are taken, check to see if nearby objects could affect the CB antenna signal. When checking a mobile rig, for example, all car doors should be closed, windshield wipers placed in the down position, and the trunk lid closed. The car should not be parked next to trees or dense foliage. The immediate surrounding area should be free of overhead power lines and the metal downspouts of a nearby house.

Keep the meter's whip antenna in a perfectly vertical position while taking readings. Important, too, is the distance between meter and antenna; the farther, the better. This prevents the meter from picking up the undesired magnetic field from the antenna. You're interested only in the electrostatic field—that part of the pattern that travels beyond the immediate area. (If the meter has low sensitivity, the type without a transistor amplifier, it usually can be improved by increasing its whip length with additional wire.) A spacing that gives good results is approximately 18 feet between meter and car antenna. Using a meter with poor sensitivity, however, might require a somewhat closer range.

Whatever type of field-strength meter is used, its whip antenna must not be 9 feet in length. This length would cause it to act as a beam element and absorb an undesirable amount of energy from the CB antenna. Then, as the meter whip retransmits this energy, it distorts the true pattern of the CB antenna. The meter whip should be at least one foot longer or shorter than 9 feet.

Finally, the meter should not be handheld, if possible. Placing it on some solid object, such as a chair, will keep the reading steady and easy to view.

# 90

## Less TV Buzz

Television sets near CB transceivers sometimes cause a raspy, buzz-type interference in the CB speaker. The source is harmonics from the television's horizontal deflection system of the television, which generates frequencies into the CB band. It is most easily identified on CB sets with a continuously tunable dial. As the knob is rotated, the buzz appears at about every 15 kHz on the dial.

One cure is to increase the distance between the television and CB set, if possible. In severe cases, internal shielding of the television cabinet may be done with metal screening fastened to the cabinet sides, bottom and top (inside). All pieces of screening must make electrical contact with each other and also connect to the television chassis. Unless the installer is familiar with the shock hazard of a "hot" television chassis, and knows how to avoid it, shielding should not be attempted on a television set of an ac – dc type.

# 91

## Choosing Coaxial Cable

There are two important considerations in selecting a coaxial cable— loss and cost. By far, the most popular cable type is RG/58U. It is the least expensive, and signal loss per foot is negligible for cable runs of less than 50 feet between rig and antenna. Another favorable feature is the relatively thin diameter of RG/58U cable. It is the easiest to snake through walls or points of entry.

When the cable run is over 100 feet a cable of lower loss will improve the signal. RG/58U is rated at an approximate 3 dB loss per 100 feet, which indicates that half the signal is lost between rig and antenna. At a cost of about two and a half times that of RG/58U is the more efficient RG8/U. A 100-foot run of this cable reduces power by only approximately 1 dB, or one-tenth. For this reason, it may be worth the extra expense to install RG8/U for installations which require about 80 feet or more of cable.

These cable types are also available in a polyfoam version. Although similar in appearance to standard coax, a special plastic foam insulator (dielectric) creates somewhat less loss. Cost is about half-again as high as standard coax. RG8/U in the polyfoam version might prove attractive for exceptionally long cable runs—over 100 feet—where losses become serious.

# 92

## Marine Installation

Marine radios operating in the 2- to 3-MHz band are required to have an extensive ground system, usually a copper plate, fastened below the water line. For CB operation in a boat this is not normally required. Antennas designed for marine CB are generally one-half wavelength long, which overcomes the ground requirement. There is, however, much to recommend the use of a ground on small boats of nonmetallic construction (wood, fiberglass, etc.). It is possible for metal items on the boat to pick up and distort the CB signal or to emphasize an ignition noise problem.

The simplest kind of grounding is done by running a common ground strap to metal objects on the boat. These might include posts which support lights, other electronic devices such as a depth finder, or decorative metal strips. The material for grounding together these units should be a strip of copper more than 2 inches wide, if possible. (Copper in roll form is generally available in local building-supply stores.) Run the strip among the various metal items, including transceiver and engine, using the shortest possible run between them. There must be good metal-to-metal contact at each point.

The handy CB user might wish to install a conventional grounding system while the boat is out of the water. One approach is to install a system which is available in kit form. One such package includes two 8-foot copper tubes which fasten along either side of the boat's false keel. These tubes provide the equivalent of 14 square feet. Necessary installation hardware and instructions are included.

A home-made ground system, which might help to eliminate difficult cases of ignition noise, is made with roll copper calculated to cover at least 12 square feet of surface under the hull. The position of the strips is not critical, but try to keep at least one section near the CB set. High-grade marine screws are used to fasten the edges of the copper to the hull. It is recommended that screws be spaced at close intervals along the edges of the copper, at least every few inches. If the copper separates anywhere from the hull, use additional screws to anchor it firmly.

Bringing the ground connection into the boat is generally done with a heavy bronze bolt through the copper and hull. A solder connection is suggested between the head of the bolt and the copper. Inside the boat, nuts and washers are used to connect a strip of copper from the bronze bolt to transceiver, and other metal objects to be grounded. Standard marine techniques, such as calking and back-up blocks, should be used where the bronze bolt runs through the hull.

# 93

## Wiring Coaxial Connectors

To prevent short circuits or open connections, wire a coaxial connector as illustrated in the four steps of Fig. 93-1. Shown is a PL-259 unit, standard plug for attaching a coaxial cable to a CB transceiver. Before starting, check to see if the end of the cable is cut squarely. Then continue as follows:

Step 1. Insert the end of the cable through both the coupling ring and adapter. Notice that the *knurled end* of the coupling ring and *narrow end* of the adapter face toward the free end of the cable. Next, carefully remove $3/4$ inch of the black cable jacket. (A useful tool is a razor blade.) Just be certain that the braid is not nicked while you are cutting away the jacket.

Step 2. Fan out the braid slightly with your fingers and fold it back on the cable as shown.

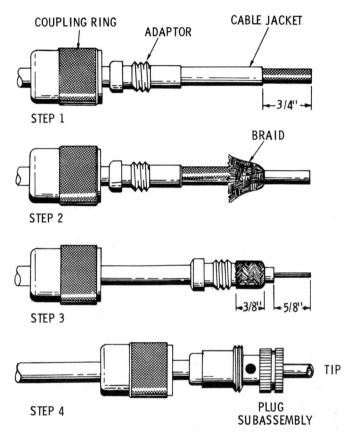

Fig. 93-1. Wiring coax connectors.

Step 3. Move the adapter under the braid and trim off the braid to an overall length of $\frac{3}{8}$ inch. Next, remove $\frac{5}{8}$ inch insulation from the center conductor. Apply a soldering iron to the center conductor and tin the wire with a bit of solder. (Be careful not to overheat and melt the insulation.)

Step 4. Take the plug subassembly and screw it on the adapter. Apply the iron at the holes and solder the braid to the assembly. Do not overheat; hold the iron at the hole for just enough time to melt the solder. After this has cooled, solder the center connector to the tip of the plug. Finally, screw the coupling ring onto the plug subassembly.

# 94

## Mobile Power Pickup

There are at least three methods for picking up power for a mobile installation—the cigar-lighter socket, ignition switch, and at the car battery. The first method, using the cigar-lighter outlet, is considered as a temporary hookup. Though it is convenient, it prevents use of the lighter and it leaves a dangling wire at the dashboard. If it is used for any considerable length of time, be aware that many lighter sockets continue to have power even after the ignition key is removed. Thus, if a rig is inadvertently left on overnight, it could discharge the battery. This would happen in the situation of a tube-powered transceiver, which draws upwards of 50 watts.

The safest power-pickup point is at the ignition switch. By hooking the power lead to the accessory terminal on the back of the switch, the set may be operated while the engine is on or off. Removal of the key automatically turns off the transceiver. (This is the same method used with most ordinary car radios.) Locating the correct terminal at the rear of the ignition switch can be done by putting your hand behind the dashboard and feeling for the longest screw on the rear of the switch. (Since it is designed to receive additional accessory wires, the shaft of the screw is longer than other terminals.) Depending on the particular car, there may be up to four screw terminals on the switch.

Some CB manufacturers recommend that the power lead should run directly to the ungrounded or hot terminal on the car battery. Their reason concerns noise suppression. By making a direct connection to the primary power source, there is less chance that noise voltage impressed on other power wiring will reach the receiver through the power lead. This can be checked experimentally and the direct battery connection used if there is a significant drop-off in noise. It should be remembered, however, that with the direct hookup, the transceiver is not turned off with the ignition key. As mentioned earlier, this is a problem with tube-type equipment. With solid-state units, however, an accidental overnight power drain would probably not fully discharge the car battery, due to the extremely low wattage in transistor equipment.

# 95

## Calling Unknown Stations

FCC rules generally forbid transmitting to a station whose call sign you do not know. There are, however, certain exceptions where it is permissible to send out a general call requesting that any listening station respond, even when call signs are not known. The most important instance is when in distress. This is defined as a time when there is immediate danger to life or property. After the emergency is over, the FCC requires that it be notified of the details. Notification should be sent to both your local FCC field office and to FCC headquarters in Washington, D.C.

Another exception to the general rules is when travelling in unfamiliar territory. It is permissible to use a CB mobile rig to contact any local operator for assistance on Channel 9. You may obtain, by calling anyone who will answer, such information as road instructions, where lodging is available, etc.

# 96

## Range Formula

Predicting the distance covered by a CB system is difficult due to the numerous factors which affect the signal. Raising the antenna height, for example, increases distance, as does using an antenna with higher gain. Distance generally improves over terrain which is flat and free of obstructions. The seasons also can affect range; the absence of foliage in winter tends to improve conditions. Transmitting over water generally increases range by approximately three times.

An influence on the radio signal which doesn't change is the curve of the earth's surface. Its effect is to limit range, since the 27-MHz signal tends to travel in straight lines outward from the antenna. Since there is virtually no bending of the signal as it reaches the horizon, it continues to travel into space. This is in contrast to standard a-m broadcast signals which curve around the earth's surface for great distances.

Since the earth's curvature is unchanging, it is possible to calculate its effect on the CB signal. All we need to know is the height of the antennas at the transmitting and receiving points. The calculation is based on the fact that the higher the antenna, the more opportunity the signal has to "look over" the horizon. The formula shown below cannot take into account various local conditions (obstructions, etc.), but it can help suggest what kind of improvement you can expect by raising the antenna system. In general, small height increases of a few feet provide range increases of only about a mile or so. Moving the antenna to a hilltop site, on the other hand, might be worth the additional expense and trouble if range is critical. The formula indicates the kind of mileage difference you can expect.

$$\text{Range in miles} = 1.23 \left( \sqrt{h_t} + \sqrt{h_r} \right)$$

where,

$h_t$ is the height of the transmitting antennas in feet,
$h_r$ is the height of the receiving antenna in feet,
1.23 is a constant.

Let's calculate an example where a transmitting antenna is 64 feet off the ground, and the receiving antenna is 16 feet high. These heights are measured from ground to the base of the antenna (where radiation is strongest).

$$\begin{aligned}
\text{Range} &= 1.23 \left( \sqrt{64} + \sqrt{16} \right) \\
&= 1.23 \, (8 + 4) \\
&= 1.23 \, (12) \\
&= 14.76 \text{ miles}
\end{aligned}$$

## Receiving Single Sideband

Single sideband, an efficient transmitting system that greatly increases power, is widely used in military, commercial, and amateur equipment. It is also approved by FCC regulations for use in CB. The standard CB receiver, however, is unable to detect the sideband signal. The audio will sound completely garbled.

There are several ways to equip the receiver for sideband reception. One is to add a beat-frequency oscillator (bfo). This, however, requires considerable skill and internal modifications to the circuit. Another method is to use a signal generator or grid-dip oscillator near the receiver while it is tuned to a sideband signal. Either instrument is adjusted to produce a 27-MHz signal. By careful tuning in the 27-MHz region, it is possible to radiate a signal into a receiver that makes the sideband audio intelligible.

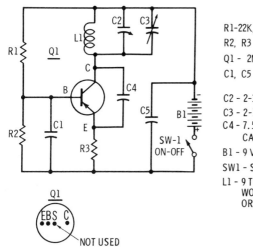

PARTS LIST

R1-22K, 1/2 WATT

R2, R3 - 3.3K 1/2 WATT

Q1 - 2N371

C1, C5 - .01 µF DISC CERAMIC CAPACITOR

C2 - 2-30 pF TRIMMER CAPACITOR

C3 - 2-15 pF VARIABLE CAPACITOR

C4 - 7.5 pF DISC OR MICA CAPACITOR

B1 - 9 VOLT BATTERY

SW1 - SWITCH, SPST

L1 - 9 TURNS NO. 26 ENAMEL WIRE, WOUND ON A 1/2" COIL FORM OR DOWEL

**Fig. 97-1. Single-sideband receiving adapter.**

If none of these methods is conveniently available, a simple 27-MHz transistor oscillator may be inexpensively constructed according to the values shown in Fig. 97-1. It is tested by tuning its capacitor while the device is near the CB transceiver. It should be possible to locate the oscillator signal by observing the receiver S-meter. While receiving a sideband signal, the oscillator is tuned back and forth over the channel frequency until the audio clears. Also adjust the strength of the oscillator signal for best results by moving the device different distances from the CB transmitter.

The device can be constructed on a small piece of perforated plastic board. Note that one lead of transistor Q1 (the S terminal) is not used and is clipped off close to the case. The values of capacitors C2 and C3 are approximate and you can use any capacitors within a few picofarads of the ratings given. C2 is a standard mica compression-trimmer type. C3 is a variable and should have a ¼-inch shaft to which a plastic knob is attached.

After the circuit is wired, turn C3 so its plates are halfway meshed. Turn on power switch SW1, and turn on the CB receiver and tune it to a channel near the center of the band. (Watch the S-meter for the signal.) This procedure sets the approximate tuning range of the oscillator. Now the knob of variable capacitor C3 is used to finely tune in sideband signals on any CB channel.

## Dust-Free Display

According to FCC requirements, a CB license must be displayed at the operating location. And it is often desired to mount QSL cards on the station wall. To preserve the appearance of these items, they should be mounted in transparent protective covering.

An inexpensive technique is to use clear plastic wrap, the type used in the kitchen for preserving food. Although it is transparent, the plastic does have the disadvantage of wrinkling. This is easily overcome by placing a cardboard backing on the paper to be protected, then folding the plastic around the card's edges. The plastic may then be retained by some adhesive tape.

## 10 Code

An informal code has come into use among CB operators for the purpose of increasing both speed of transmission and intelligibility. By memorizing the 10-code numbers and their meanings, it is possible to shorten otherwise long sentences. Since the code is not an official requirement, be sure that the other operator is aware of 10-code meanings before using them.

| Code | Meaning |
|------|---------|
| 10-1 | Receiving poorly. |
| 10-2 | Receiving well. |
| 10-3 | Stop transmitting. |
| 10-4 | O.K. |
| 10-5 | Relay message. |
| 10-6 | Busy. |
| 10-7 | Out of service—leaving the air. |
| 10-8 | In service—subject to call. |
| 10-9 | Repeat, reception bad. |
| 10-10 | Transmission completed, subject to call. |
| 10-11 | Talking too rapidly. |
| 10-12 | Officials or visitors present. |
| 10-13 | Advise weather and road conditions. |
| 10-18 | Engineering test. |
| 10-20 | What is your location? |
| 10-21 | Call . . . station by phone. |
| 10-23 | Stand by. |
| 10-24 | Trouble at station. |
| 10-25 | Do you have contact with . . . ? |
| 10-30 | Does not conform with rules and regulations. |
| 10-33 | Emergency traffic at this station. |
| 10-36 | Correct time. |
| 10-65 | Clear for message. |
| 10-92 | Your quality is poor. |
| 10-99 | Unable to receive your signals. |

## SECTION 6

# Additional Information

# H.E.L.P.

Some 1500 CB groups are, at the time of this writing, members of H.E.L.P., which stands for *Highway Emergency Locating Plan.* The program is intended to provide a nation-wide network of monitoring stations for assisting the motorist in distress. Aside from organizing CB stations already on the air, the program seeks to encourage use of CB equipment in every automobile. The emphasis is on inexpensive equipment for the motorist which would be minimally capable of operation on Channel 9, or any other frequency set aside for H.E.L.P. operation.

The program was initiated by the Automobile Manufacturers Association. According to this organization, conventional methods of assisting stranded motorists have failed to fully provide essential services. These include roadside telephones and increased patrolling of roads by police. In the H.E.L.P. program a motorist in trouble transmits a call over Channel 9, which is picked by a monitoring station. The monitor either sends the required assistance or relays the call to the appropriate agency.

Details on how to participate in the H.E.L.P. program may be obtained by writing to: Automobile Manufacturers Association, 320 New Center Building, Detroit, Michigan.

## React Program

Standing for *Radio Emergency Associated Citizens Teams,* REACT is an unofficial nation-wide organization of CB monitoring stations. Its purpose is public service and its goal is to provide 24-hour-a-day listening for emergency and service calls throughout the United States. The calling channel is generally accepted as Channel 9 on the Citizens band. Monitoring stations, who serve on a voluntary basis, will respond to a call for assistance, or provide travel instructions.

The REACT program was originated by a major CB manufacturer. There is no membership fee and the sponsoring company will provide such items as an automobile sticker, membership card, and news bulletins to interested organizations. Details on the program may be obtained from: 111 E. Wacker Drive, Chicago IL 60601.

# Channel Use

The 23-channel Citizens band is shown on the following page with its assigned frequencies. How the channels may be employed is shown under the column "Use by _____ Stations." Where the word "Same" appears, it refers to permissible use by units of the *same* station, all operating under the same call sign that is, a base and its mobiles. (This is also termed *intrastation* activity.) Such traffic by "Same" stations may be conducted on any of the 23 channels.

The entry "different" for a particular channel refers to activity between stations of *different* call sign, or interstation communication. This is limited to the seven channels indicated. The only exception occurs in time of danger to life or property.

It is usually to the advantage of stations under the same call sign to operate on channels other than 10 through 15 and 23, where interference from interstation activity tends to be greatest.

Note that a unit operated under Part 15 of FCC regulations may operate on all Class-D channels except Channel 1. Thus it is technically possible for a Class-D station to communicate with a Part 15 unit. This operation, however, is prohibited unless the following qualification is met: The 100 milliwatt unit must also be licensed under Part 95 (Class D) of FCC regulations and be operated in accordance with its provisions.

Interweaved with 23 Class-D channels are six Class-C Citizens band channels. In most cases, the frequencies are not the same as Class D, but are close. The Class-D station will not normally receive interference from Class-C channels since they are used mostly for short-duration signals for controlling model aircraft. In fact the reverse has been true: model aircraft have gone out of control due to interfering Class-D transmissions. This has led to a granting of new channel space in the 70-MHz band for radio-control enthusiasts.

On Channel 23 some interference from other services may be expected, especially remote control of commercial equipment. In some areas, the channel contains tone signals which are constantly repeating. Monitoring of Channel 23 in your area should indicate whether it is usable or not.

In some areas, local groups have set aside channels for special use. Although the practice is unofficial, the FCC does not prohibit such activity.

| Channel | Frequency (MHz) | Used by _____ Stations | Remarks |
|---|---|---|---|
| 1 | 26.965 | same/different | Not for Part 15, use |
| 2 | 26.975 | same/different | |
| 3 | 26.985 | same/different | Class C; 26.995 |
| 4 | 27.005 | same/different | |
| 5 | 27.015 | same/different | |
| 6 | 27.025 | same/different | |
| 7 | 27.035 | same/different | Class C; 27.045 |
| 8 | 27.055 | same/different | |
| 9 | 27.065 | emergency and motorist assistance | |
| 10 | 27.075 | same/different | |
| 11 | 27.085 | calling channel only | Class C; 27.095 |
| 12 | 27.105 | same/different | |
| 13 | 27.115 | same/different | Marine channel (unofficial) |
| 14 | 27.125 | same/different | |
| 15 | 27.135 | same/different | Class C; 27.145 |
| 16 | 27.155 | same/different | SSB (unofficial) |
| 17 | 27.165 | same/different | |
| 18 | 27.175 | same/different | |
| 19 | 27.185 | same/different | Class C; 27.195 |
| 20 | 27.205 | same/different | |
| 21 | 27.215 | same/different | |
| 22 | 27.225 | same/different | |
| 23 | 27.255 | same/different | Shared with Class C and other services |

## Speaker Check

A simple speaker test, using no special equipment, can be done with a flashlight battery. Any size 1.5-volt cell may be used. Hook the battery across terminals (either way). If the speaker is in good condition, a loud "click" should be heard as wires from the battery make contact with the speaker terminals.

# Official Sources

Following is a list of government sources and addresses incident to the CB field. Applications for a CB license or modification can be obtained from the Washington, DC 20554 office of the commission, or from the nearest FCC field engineering office (see below). All applications are mailed (with $4 filing fee) to:

Secretary
Federal Communications Commission
P.O. Box 1010
Gettysburg, PA 17325

Requests for copies of the FCC Rules & Regulations (you are required to have Volume VI) are addressed to:

Superintendent of Documents
U.S. Government Printing Office
Washington, D.C. 20402

### FIELD ENGINEERING OFFICES

Address all communications to Engineer in Charge, FCC.
Alabama, Mobile 36602
Alaska, Anchorage (P.O. Box 644) 99501
California, Los Angeles 90012
California, San Diego 92101
California, San Francisco 94111
California, San Pedro 90731
Colorado, Denver 80202
District of Columbia, Washington 20554
Florida, Miami 33130
Florida, Tampa 33602
Georgia, Atlanta 30303
Georgia, Savannah (P.O. Box 8004) 31402
Hawaii, Honolulu 96808
Illinois, Chicago 60604
Louisiana, New Orleans 70130
Maryland, Baltimore 21202
Massachusetts, Boston 02109
Michigan, Detroit 48226
Minnesota, St. Paul 55101
Missouri, Kansas City 64105
New York, Buffalo 14203

New York, New York 10014
Oregon, Portland 97204
Pennsylvania, Philadelphia 19106
Puerto Rico, San Juan (P.O. Box 2987) 00903
Texas, Beaumont 77701
Texas, Dallas 75202
Texas, Houston 77002
Virginia, Norfolk 23510
Washington, Seattle 98104

## CB Club Activity

One way to add interest to a CB club meeting is to show movies. There are hundreds of free films available for group showing, with many titles in electronics and similar fields of CB interest. You might begin by inquiring at the local power and telephone company. Public utilities frequently offer films on a loan basis. A rich source of free films is the U.S. Government. You may obtain further information from: Superintendent of Documents, U.S. Government Printing Office, Washington, D.C. 20402. These films are made available only to clubs and organizations, rather than to the individual.

## Commenting on Proposed Rules

From time to time Citizens band regulations are changed and amended by the FCC. But before they become law, any interested party may file comments on them. The Commission has stated that it relies on these suggestions from the general public for guidance. There are, however, certain practices which help make comments carry maximum weight.

A simple "for" or "against" letter to the FCC about a given proposal is apt to have little influence. The Commission requires a *constructive* reason why the rule should be changed. For example, merely stating that you *disagree* with the 20-foot rule on antennas is not persuasive. But arguing that a certain antenna design, by rising to 22 feet can eliminate tvi, the comment could merit considerable attention.

Comments should be submitted in typewritten form. The FCC requests that fourteen copies be included so they may be distributed among its various offices.

# Common Violations

Through its monitoring facilities, the FCC checks on CB operation for rules violations. In cases of serious violations fines are levied and licenses are revoked. Some common violations include:

*Unnecessary communications.* These are usually messages not related to the operator's business or personal activities.

*Off-frequency operation.* This refers to a condition where the transmitter is beyond the specified 0.005-percent frequency tolerance. This should be checked occasionally by a licensed technician equipped with an accurate frequency meter.

*Communication beyond ground wave.* Contacts may not be made with stations located farther than 150 miles away.

*Improper station identification.* This includes dropping prefixes or failure to give call letters.

# Accident Scene

What to do at the scene of an accident? Since a CB-equipped car is in a special position to summon help at roadside emergencies, here are tips on how to assist effectively.

According to one experienced state trooper, a big problem is that many people don't know how to describe the location of the accident. They might say something like, "Two miles south of Route 10." "But most people," says the officer, "simply can't judge mileage accurately enough." The result is that police, ambulance, or other help uses extra, often precious, minutes to find the accident scene.

The recommended way for identifying location is to describe the nearest intersection or landmark. A business sign might also prove helpful. Don't rely only on the route number and approximate position. One way to pinpoint the location, especially if it's off the road and not clearly visible, is to judge distance by observing nearby telephone poles. Since poles are usually about 150 to 200 feet apart, they can help gauge distance from a nearby landmark.

Another suggestion is how to minimize the possibility of fire in a car wreck. If you carry a common pair of slip-joint pliers in your car, it enables you to quickly unscrew one cable from the car battery. Loss of electrical power reduces the danger of a hot wire causing fire.

Flares are useful for waving traffic safely around the accident scene. But it's also important to keep flares well away from spilled gasoline, often found at a car wreck. A final tip: while assisting at an accident never turn your back to oncoming traffic. And while you are directing traffic, take up a position well in advance of the accident scene.

## Reading Faded Tube Numbers

On removing a tube from an older transceiver for checking, you may discover that the tube-type number has faded. There are three possibilities for making the numbers readable. Slowly turn the tube under a strong light; the numbers might become visible at some point. If this doesn't work, blow your breath on the tube. The mist that forms on the glass may make the numbers appear. Finally, place the tube in a refrigerator. After the glass is chilled, it may then be possible to view the numbers.

## Sporadic-E Interference

A CB transceiver situated close to a television set can produce interference to sound and picture. (How to cope with these problems is discussed elsewhere in this book.) There is, however, a type of television interference that may mislead a television viewer into believing that the origin of the interference is a nearby CB set. Understanding the nature of the problem could help to counter a false accusation.

Especially during spring and summer, there appears high in the earth's atmosphere a mass of electrified gas. It is termed the *sporadic-E layer*. It is actually a portion of the ionosphere, a region beginning some 50 miles up, which has the ability to produce long-distance signal-skipping on frequencies up to about 30 MHz. But the sporadic-E layer differs from the conventional ionosphere in that it causes long-distance transmission of vhf signals above 30 MHz, where television stations are assigned. Due to sporadic-E transmission, it is common for television signals during warmer months to skip as far as 1,000 miles. The frequent result is co-channel interference. A distant television station on a given channel may arrive and interfere with a local television station on the same channel.

Certain clues identify sporadic-E interference. One is that the effect is not constant; these electrical clouds tend to shift quickly, even disappear and return. The basic pattern, however, is that an interference peak generally happens over several days at a time during spring and summer months. Also, the effect tends to be strongest for several hours after sunrise and several hours after sunset (during prime television viewing times).

There is also a particular pattern of interference viewed on the television screen and heard in the speaker. In severe cases, it may be possible to see the image of the distant station faintly on the screen. Sometimes it fails to lock in or synchronize properly, and appears to move rapidly in a series of frames behind the picture of the strong local station.

Television audio is affected in a characteristic manner. As the two signals —from local and distant station—mix in the television set they produce an audible signal. It is generally a high-pitched, ragged noise.

Since sporadic-E interference affects all television reception in a given area, you might inform a complaining neighbor to check with friends in the same town for similiar interference. Another convincing bit of evidence— the interference should be on the neighbor's screen even as he speaks to you, and your CB set is turned off.

# Roof-top Safety

Erecting an antenna on a steep roof can be a hazardous operation. To minimize the danger, here are several tips that have proved useful in antenna installation. The first is to complete as much of the assembly as possible on the ground. This allows the completed antenna to be hoisted or carried to its mounting point with little need for swinging tools and elements while in a precarious roof-top position.

The most serious threats during antenna installation are falling and electrocution. The use of a rope can greatly reduce the risk of falling from the roof. Tie down one end securely at or near the ground, then toss the other end over the peak of the roof. The rope should cross over near where the antenna is to be mounted. Next, tie down the rope on the other side of the house. The rope then provides a firm point for grasping as you make your way along the roof's steep slope, or if you lose your balance.

The second hazard, electrocution, is a very real threat to anyone working with antennas. In most cases it has happened while a person was grasping aluminum antenna elements while attempting to bring them into proper position. Electrical contact was made as the upper, or far, end of an element brushed against a nearby power line.

Such accidents can be prevented by mounting an antenna away from any electrical lines (which can also interfere with antenna performance). Also, the outer end of the elements should be watched as they are moved into place to avoid striking anything. Finally, it is a good idea to have someone assist during antenna installation and warn of any potentially dangerous situation.

# Index

# Index